Quantitative Elements of General Biology

Ivan Maly

Quantitative Elements of General Biology

A Dynamical Systems Approach

Springer

Ivan Maly
Department of Physiology and Biophysics
State University of New York at Buffalo
Buffalo, NY, USA

ISBN 978-3-030-79145-2 ISBN 978-3-030-79146-9 (eBook)
https://doi.org/10.1007/978-3-030-79146-9

This Springer imprint is published by the registered company Springer Nature Switzerland AG
The registered company address is: Gewerbestrasse 11, 6330 Cham, Switzerland

Contents

1 Introduction 1
References 6

2 Biomass Growth and the Language of Dynamic Systems 7
2.1 Equilibria, Stability, and Hysteresis 7
2.2 Growth with Generations: Oscillations and Chaos 10
References 14

3 Species and Speciation 17
3.1 Selection-Mutation Equilibria 17
3.2 Neutral Sympatric Speciation 24
References 33

4 Signaling and Control 35
4.1 All-or-Nothing Response 35
4.2 Oscillatory Control 41
4.3 Switches 52
References 58

5 Self-Organization in the Cell 61
5.1 Orientation of the Cortex 61
5.2 Emergence of a Focal Cytoskeleton 74
5.3 Biochemistry of Cell Polarity 87
References 96

6 Forces that Shape the Cell 99
6.1 Symmetry-Breaking and Emergent Irreversibility 99
6.2 Cell Division and Alternative Equilibria 112
6.3 Cell Interaction and Multiperiodic Motion 129
References 139

7 Multicellular Morphogenesis 143
7.1 Turing Mechanisms 143
7.2 Phyllotaxis 147
7.3 Segmentation 153
References 158

8 Interspecies Dynamics 161
8.1 Coexistence and Exclusion 161
8.2 Temporal Patterns 164
8.3 Spatial Effects 170
References 174

9 Summary and Outlook 177
References 183

Appendix: Phase Plane Analysis 185

References 189

Index 191

Abbreviations

cAMP	Cyclic adenosine monophosphate
CAP	Catabolite activator protein
EGF	Epidermal growth factor
EGFR	Epidermal growth factor receptor
FGF	Fibroblast growth factor
MAPK	Mitogen-activated protein kinase
MAPKK	Mitogen-activated protein kinase kinase
MAPKKK	Mitogen-activated protein kinase kinase kinase
RTN	Retrotrapezoid nucleus
WASP	Wiskott-Aldrich syndrome protein

Chapter 1
Introduction

As a field of study, general biology seeks to identify dynamic processes and functional structures that are common to all life forms. Owing to the extreme diversity of living organisms, the goal of general biology can also include identification of what is sufficiently widespread among life forms, and explanation of the mechanistic reasons for the variation. General biology is therefore related in its methodology to comparative biology, although the emphasis is on commonality and not the differences. Toward this aim, general biology integrates the results of disciplines such as cell biology, physiology, ecology, and others, insofar as they are not specific to the taxa studied. The range of topics covered by general biology acquired its present breadth by the time when the fundamental findings in molecular biology became fully systematized. These topics include (see Libbert 1982) the distinction of living from non-living matter, metabolism, genetics, mechanisms of proliferation, embryology, control mechanisms, behavior, evolution, and ecological interactions. In addition to a dramatic progress that has been achieved in many of these areas with the application of novel methods in recent years, a new emphasis on system-level properties that are common to different taxa has emerged (see, *e.g.*, Alon 2006; Maly 2008; Klipp et al. 2009; Prokop and Csukas 2013; Rajewsky et al. 2018).

The quantitative approach has been indispensable to advancement of general biology. Beginning with the example of Mendel, development of statistics was intertwined with the progress in genetic research (see, for example, Fisher 1999), a process that continues with today's machine learning and genomics. To take an obvious and well-studied example, whether one, two, three, or four leaves are formed at a time during the growth of a plant shoot is a quantitative question that we are to have a systematic answer for if general understanding of plant habits is the goal. However, development of self-organization science in the twentieth century (*e.g.*, Haken 1983) has shown that the more fundamental problem of why a defined number of discrete organs (leaves) is formed in such cases, instead of a random number of them or, say, a continuous collar of light-capturing tissue, is also a problem to be answered through application of quantitative methods. In this example as

I. Maly, *Quantitative Elements of General Biology*,
https://doi.org/10.1007/978-3-030-79146-9_1

in many others of its nature, the emergence of the biologically functional pattern is a quantitative effect arising in a system of interacting elements (cells, molecules) and cannot be explained—whether qualitatively or with precision—other than as a property of the dynamics of the system components' interaction. Although application of the methods of dynamical systems research to biology has a considerable pedigree (see, *e.g.*, Rosen 1970) and has impacted the development of the mathematical methods themselves (Tu 1994; Jost 2005), the recent emphasis on systems explanations in biological science has led to a solidification of the dynamical systems approach (*e.g.*, Murray 2002; Garfinkel et al. 2017; Chen and Zaikin 2021).

The quest for identification of processes general to all life has suffered from a problem threatening to reduce all answers to something very near tautology: since reproduction and inheritance are attributes that unquestionably belong to the definition of life, and we accept all life—whether extant or known from the paleontological record—to have common ancestry, any features identified as common may simply be inherited. Indeed, this could be so whether these features be deterministically attached to all life or merely accidents of life's origin and early evolution. While this problem may not be fully resolved for some time, advances in astrobiology and particularly in methods of detection and future study of life on other planets (Yamagishi et al. 2019) are beginning to supply hope that juxtaposition of life as we know it with independent, or at least widely separated life forms may become possible. Such comparison would be uniquely informative for general biology, as it would enable separating the accidental, masquerading as universal, from what is truly universal among the properties of living matter.

For general biology, therefore, this is the time to systematize our understanding from the perspective of identifying, among the features common to life as we know it, those that are likely generalizable beyond the singular example we have been studying. Even if at present this may appear to be an exercise in hypothesis-making, demand for a well-grounded set of generalizable life attributes can become acute very soon with the accelerating pace of developments in astrobiology. As regards the generalizable planetographic conditions for life's emergence and existence, as well as planetographic signatures of extant life, this direction of study is comparatively advanced (Yamagishi et al. 2019). What will be required following a definitive discovery of extraterrestrial life—which we expect to come in the form of discovery of its planetographic signature—will be guidance from general biology with respect to the dynamic and structural features of such life that may be rationally expected, based on our experience with studying life terrestrial, so that more detailed data could be efficiently collected and interpreted.

General biology, now understood more narrowly as generalizable biology in the above sense, can, in principle, encompass a very wide range of life attributes. For example, requirements concerning the chemical composition of living matter may be addressed. Not even a review of possible directions in such a broad, interdisciplinary study will be attempted here. Instead, our subject will be confined to sketching out a set of dynamical attributes that is identifiable with life and can be expected to be generalizable from life as we know it to its possibly unrelated forms that may be discovered in the future.[1]

Having considered the old and new motivation for identifying the general (*i.e.*, generalizable) dynamical systems properties and processes that are associated with living matter, how can we approach this problem on our present level of understanding and in a format that is both writable and readable? A compendium (the word from the title of the first work cited in this chapter, a classic of non-mathematical general biology) of dynamical systems results pertinent to generalizable biology is not feasible at this point, before the overall direction has been sketched out, debated, and accepted. What follows, therefore, is of necessity an incomplete and somewhat subjective version of such a sketch, but one that is intended to serve the modest purpose just stated.

When selecting the topics to cover and their sequence, we need to be mindful of the problem of circularity in all biological reasoning, which stems from the already mentioned issue of the subject of our study arising from self-replication. A process may be mechanistically fundamental with respect to another process, yet its very existence may be contingent on the higher-level process and its kinetics controlled by the latter through evolutionary dynamics, whereby fundamental processes are fine-tuned to answer the needs of the higher-level ones. Today's didactic literature covering general biology comports with the paradigm of the cited twentieth-century compendium and usually adopts the mechanistic logic, proceeding from the molecular basis of life, through physiology, to ecology and biogeography. Here, we will take a slightly different approach. Processes responsible for the very existence of living matter in defined forms will be considered first (ecologically abstracted biomass growth kinetics, followed by elementary speciation and homeostasis). From there on, our sequence will be more conventional, covering elements of cell biology, multicellular development, ecology, ending in treatment of evolution from the standpoint of species interaction. A degree of circularity in the exposition, whereby the most elementary ecology and evolution is considered first, and more complex effects in the same realms last, should be seen as intentionally reflecting the mentioned circularity in the subject matter. With this caveat, the exposition is from the elementary to the complex, and from the foundational to the mechanistically derivative.

An additional reason—besides its being fundamental to the very existence of living matter—to consider biomass growth first is that useful formalisms exist in this area of study that are both simple and illustrative of the basic concepts of biological kinetics and analysis of systems behavior. Indeed, by introducing these on the example of elementary, single-species biomass kinetics, we remain in the mold of the didactic tradition exemplified by the cited work of Murray. The need for such an opening chapter arises from our desire to make this monograph accessible also to those readers who may be less familiar with the language of dynamical systems

[1] Since this was written, a possible phosphine biosignature has been discovered on Venus and an expedition for sample collection and study in situ has been proposed (Greaves et al. 2020).

theory. In addition to covering its biological subject, the biomass kinetics chapter (Chap. 2) can serve as a technical introduction to the quantitative approach that is followed in the rest of the book. Both continuous-time models and models with generations will be considered, the former introducing the concepts of steady states and stability analysis, and the latter, emergence of oscillations, multiperiodicity, and apparent chaos.

The jump to speciation in the third chapter may be unexpected, yet existence of species is fundamental to any quantitative treatment of organismal mechanisms and therefore precedes them in our exposition. The concept of species as defined forms that living matter can take supplies the basis for having well-defined parameter values in any model. All subsequent material in the book relies on existence of a mechanistic structure and a set of biophysical parameters that are representative of some identifiable quantity of living matter. This is possible only when the living matter is subdivided into at least tolerably delineated varieties. With this in mind, we will consider (Chap. 3) the most elementary processes in evolution, which can lead to emergence and maintenance of species. The basic results of quasi-species theory of mutation-selection balance will be laid out, followed by analysis of neutral sympatric speciation in a spatially well-mixed population.

Equally foundational to existence of all living matter are control processes that can maintain the internal environment of organisms in the variable external environment. Physiological adaptation to changing external conditions also belongs to this category of processes, as are the capabilities of organisms and cells to effect transitions between distinct physiological states when survival requires it. On the organism as well as cell level, processing of external signals is intertwined with the control processes, and the processing and control functionalities often share the same molecular substrate and rely on closely related types of nonlinear dynamics. In view of this, selected elementary examples of signal processing and functional control will be treated together in Chap. 4, to include all-or-nothing response, oscillatory control, and kinetic switches on the genetic, biochemical, cellular, and organism level.

The functional structure of the cell is the central biological element that connects life's molecular building blocks with the organismal structure and function. Owing to this, two chapters will be devoted to general biology problems that concern dynamical origins of the cell structure. In the first one (Chap. 5), we will deal with molecular self-organization in the cell. An important aspect of these processes that distinguishes them from molecular self-organization that may take place outside the biological realm is the orchestration of the molecular processes by the constraints imposed by the heritable cell structure (Harold 2001). How the large-scale structure is shaped by the kinetics of interaction of its molecular components will be addressed through the prism of quantitative models of cytoskeleton dynamics, cytoskeleton-directed and -driven motility, and spatially distributed cell signaling. Selected approaches to the fundamental problems of centered organization of advanced (eukaryotic) cells, orientational order of their cytoskeletal filaments, and overall polarity (directionality) of the cell structure will be highlighted. In Chap. 6, we will transition to treatment of system-level effects arising from mechanical forces that are inherent in the elementary cell structure and that shape this structure further. Here—in the belief that, like the cellularity itself, the basic cytoskeletal organization should be generalizable from the known advanced life forms—the focus will be on the microtubule cytoskeleton, whose mechanics will be analyzed in the situations of isolated interphase cells, cell division, and cell-cell interactions.

The mechanisms responsible for generation of the body plan of advanced organisms have long been recognized among the central problems in general biology. The problem of multicellular morphogenesis motivated Turing in was to become one of the pivotal achievements of the twentieth century science: demonstrating a mechanism for emergence of stable spatial patterns in an initially homogeneous molecular system (Turing 1952). In Chap. 7, we will examine multicellular morphogenesis on two examples, phyllotaxis and segmentation. Phyllotaxis—developmental arrangement of leaves and other organs of vascular plants—is an example of morphogenesis that is in evidence in the world around us. Although its mechanistic substrate is widely different, as a process whereby morphological elements of the organism are produced repeatedly and regularly, phyllotaxis can be seen as related closely to segmentation in animal body plans. Outlines of our current understanding of the two comparatively complex and well-studied processes will serve to illustrate dynamical systems principles that may be broadly operational in multicellular morphogenesis.

Having sketched out the selected elements of general biology from the molecular processes to the multicellular organism form, in Chap. 8 we return to the foundational biological processes of biomass growth kinetics and evolution to reconsider them on the level of species interaction. The focus this time will be on accounting explicitly for the ecological relationships (*e.g.*, predator-prey, mutualism, or competition) as they influence the species' population dynamics and evolutionary trajectory. The ecological factors have temporal, spatial, and genome-space aspects, and may lead to self-organization in each of the three domains. The biomass dynamics considered here will be a multi-species generalization of the elementary kinetic laws from Chap. 2, and the ecological dynamics in the genotype space, derivative from the elementary formalizations for speciation given in Chap. 3. The oscillatory temporal pattern that can arise in a simple ecological model of Lotka and Volterra will be considered fist. In the remainder of the chapter, more complex, and apparently aperiodic, temporal pattern of punctuated equilibrium in an explicitly genetic model of ecological coevolution will be analyzed, followed by a discussion of spatial dynamics associated with various ecological relationships and speciation in spatially distributed populations.

The breadth of the presented topics demonstrates unity of dynamical laws and analytical approaches across several levels of biological organization, which has been attained through recent research in the different fields. While much work remains to be done to explicitly unify biological theory on the basis of the mathematical apparatus of dynamical systems, the very rough sketch of such an encompassing theory that is offered here may serve as an additional impetus for such future effort.

References

Alon U (2006) An introduction to systems biology: design principles of biological circuits. CRC Press, Boca Raton, FL

Chen S, Zaikin A (2021) Quantitative physiology. Springer, Singapore

Fisher RA (1999) The genetical theory of natural selection. Oxford University Press, Oxford

Garfinkel A, Shevtsov J, Guo Y (2017) Modeling life: the mathematics of biological systems. Springer, New York

Greaves JS, Richards AMS, Bains W et al (2020) Phosphine gas in the cloud decks of Venus. Nat Astron. https://doi.org/10.1038/s41550-020-1174-4

Haken H (1983) Synergetics: an introduction. Springer, Heidelberg

Harold FM (2001) The way of the cell. Oxford University Press, New York

Jost J (2005) Dynamical systems: examples of complex behaviour. Springer, Berlin

Klipp E, Liebermeister W, Wierling C et al (2009) Systems biology: a textbook. Wiley, Weinheim

Libbert E (ed) (1982) Kompendium der allgemeinen Biologie. Gustav Fischer Verlag, Stuttgart

Maly IV (ed) (2008) Systems biology. Springer, New York

Murray JD (2002) Mathematical biology I: an introduction. Springer, New York

Prokop A, Csukas B (2013) Systems biology: integrative biology and simulation tools. Springer, Dordrecht

Rajewsky N, Jurga S, Barciszewski J (eds) (2018) Systems biology. Springer, New York

Rosen R (1970) Dynamical system theory in biology: stability theory and its applications. Wiley-Interscience, New York

Tu PNV (1994) Dynamical systems: an introduction with application to economics and biology. Springer, Berlin

Turing AM (1952) The chemical basis of morphogenesis. Philos Trans R Soc Lond B 237:37–72

Yamagishi A, Kakegawa T, Usui T (2019) Astrobiology: from the origins of life to the search for extraterrestrial intelligence. Springer, Singapore

Chapter 2
Biomass Growth and the Language of Dynamic Systems

2.1 Equilibria, Stability, and Hysteresis

Dynamics of biomass growth is fundamental to the very existence of living matter. Its models (see, e.g., Hastings 2000; Kot 2001) also allow us to introduce the basic concepts of biological kinetics and analysis of systems behavior on mathematically simple examples (e.g., Edelstein-Keshet 1988; Murray 2002; Riznichenko 2011). Consider the description of the biomass of a certain population with a single variable, x. A series of instructive cases can be captured by a general form in which the rate of change with time t is a polynomial in x that extends only to the third power.

The simplest case is

$$\frac{dx}{dt} = \left(k_g - k_d\right)x, \tag{2.1}$$

where k_g is the first-order rate constant of growth and k_d, of die-off. The solution to Eq. (2.1) is an exponential function, and the population growth exhibits different properties, depending on the relationship between k_g and k_d. If $k_g > k_d$ and x is positive, the biomass increases in an accelerated and unlimited fashion. The latter feature is, of course, nonphysical. When $x = 0$, no dynamics takes place. However, any positive value of the biomass, no matter how small, triggers the runaway growth. We conclude that the equilibrium point $x = 0$ is unstable. Biologically, it means that any number of organisms colonizing the space under consideration can successfully seed a new population. This is frequently the case with simple organisms under favorable growth conditions, *e.g.*, when bacteria are inoculated into warm milk. If, on the contrary, $k_g < k_d$ (and x is still positive), then the biomass shrinks and approaches zero in the limit. This case represents an environment that is unfavorable to the species in question. As before, no dynamics takes place when $x = 0$, but since any positive biomass returns to zero with time, this equilibrium point is now stable with respect to perturbations.

I. Maly, *Quantitative Elements of General Biology*,
https://doi.org/10.1007/978-3-030-79146-9_2

Consider now the case where

$$\frac{dx}{dt} = -k_d x + k_g x^2, \tag{2.2}$$

$k_d > 0$, and k_g—also positive—is redefined as a second-order rate constant of growth. Whereas the last example was applicable to asexually reproducing organisms (or self-fertilizing hermaphrodites), the case represented by Eq. (2.2) can describe a population reproducing sexually. Since our description remains univariate, it would be appropriate for a population that is well-mixed and has a fixed ratio of sexes or, alternatively, one that consists of obligatorily cross-fertilizing hermaphrodites. Such organisms need to meet to reproduce, and the rate at which organisms meet pairwise can be thought of as being proportional to the second power of the organisms' spatial density (analogously to second-order chemical kinetics reflecting pairwise molecular collisions). Population effects that may arise from spatial variation will be left for a later chapter, and reproduction with defined generations will be considered immediately following this introductory exposition. In the absence of any spatial variation and any consequential age structure, the density of organisms is simply proportional to the population's total biomass, justifying the simple model behind Eq. (2.2).

The properties of the dynamics described by Eq. (2.2) can be deduced by examining the equation's right-hand side. This is a parabolic function of x that has roots at 0 and k_d/k_g. It is negative between these roots and positive elsewhere. Accordingly, the biomass is driven to zero from any value that is smaller than k_d/k_g, and the equilibrium point $x = 0$ is stable. Any starting value that is above k_d/k_g leads to an unrestricted further growth. In other words, a critical seeding mass is necessary to ensure the survival of the population, which is realistic for sexually reproducing organisms. At the same time, the case of Eq. (2.2) still features the unrealistic possibility of population explosion.

The final case in our introductory series of homogeneous models is the one where

$$\frac{dx}{dt} = -k_d x + k_g x^2 - k_i x^3. \tag{2.3}$$

Here, k_i is introduced as a third-order rate constant of interference between the individuals ($k_i > 0$). The motivation for such an approach is that the population may experience negative effects of crowding, for example, competition for food. The situation captured is the one in which individuals of the population need to meet to interfere with each other. The interference effect of meeting in pairs could be taken into account by adjusting the second-order constant, already named k_g (see Eq. 2.2), which remains a constant of growth as long as the negative effect of meeting in pairs does not outweigh its positive effect on the population. To reflect this situation, we still assume that k_g is positive. The next term capturing the rate of a process whereby individuals meet in space is third-order in x, and the rate constant that reflects its effect on the population biomass is k_i. Higher-order terms may be neglected because they represent more infrequent events (simultaneous meeting of four or more individuals).

In this case (Eq. 2.3) we again have $x = 0$ as an equilibrium point. Other roots of the right-hand side (the population growth rate function) depend on the values of the rate constants. When $k_g^2 < 4k_dk_i$, there are no other equilibrium points and the rate function is negative for all $x > 0$. In other words, the population cannot be established: $x = 0$ is a stable equilibrium and any starting biomass will eventually die off. If k_g^2 happens to be exactly equal to $4k_dk_i$ (which is improbable but worth considering for completeness), there is a second root, $k_g/(2k_i)$, but the population growth rate is still negative for all $x > 0$ except this one value. The equilibrium point $x = k_g/(2k_i)$ is stable with respect to perturbations that increase x but unstable with respect to perturbations that decrease it and therefore on the whole unstable. We can see that, in essence, the properties established for $k_g^2 < 4k_dk_i$ hold in this special case as well. Finally, if $k_g^2 > 4k_dk_i$, then there are two roots in addition to $x = 0$:

$$x_1 = \frac{k_g - \sqrt{k_g^2 - 4k_dk_i}}{2k_i}, \quad x_2 = \frac{k_g + \sqrt{k_g^2 - 4k_dk_i}}{2k_i}.$$

The growth rate function is negative between 0 and x_1, positive between x_1 and x_2, and negative for $x > x_2$. Zero is therefore still a stable point, as is x_2, but x_1 is unstable. In other words, x_1 represents a critical biomass: any size that is smaller will die off, whereas any that is larger will seed a persistent population which will eventually stabilize at size x_2. This model captures the realistic properties of an isolated, sexually reproducing population of organisms. Its salient properties—whereby stable equilibrium points are separated by unstable equilibrium points, which can all be found without solving the dynamic equations themselves—are common also to more complex and multivariate systems.

Lastly, among the continuous-time univariate models, it is instructive to consider a version in which a zero-order term is present:

$$\frac{dx}{dt} = k_m - k_d x + k_g x^2 - k_i x^3. \tag{2.4}$$

The constant $k_m > 0$ can represent the rate of migration of organisms into the spatial domain where the biomass is measured—whether active or passively following a flux of the medium in which they reside. The behavior of this model can be visualized qualitatively if we note that it is analogous to the last case, except that to find its equilibria we need to consider not where the three terms dependent on x equal zero but where they equal $-k_m$:

$$-k_d x + k_g x^2 - k_i x^3 = -k_m. \tag{2.5}$$

Analogously to the roots of the rate function considered in the version without immigration, Eq. (2.5) can be satisfied at one, two, or three values of x. Interesting behavior can be observed when the response of the system to varying the migration

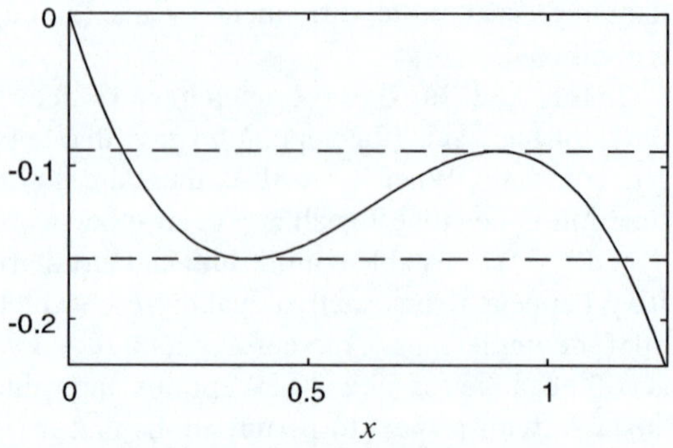

Fig. 2.1 Left-hand side (*solid line*) of Eq. (2.5). In this example, the biomass and time are unitless, $k_g = 1.9$, and $k_d = k_i = 1$. Right-hand side (*dashed line*) is plotted for $k_m = 0.089$ and $k_m = 1.605$

constant is analyzed. The conditions for such behavior arise when the left-hand-side of Eq. (2.5) is negative for all $x > 0$ and has two points where its derivative is zero (Fig. 2.1). The first of these conditions has in fact already been worked out in terms of the rate constants in the last example (amounting to $k_g^2 < 4k_dk_i$), and the second one requires that $k_g^2 > 3k_dk_i$. If k_m is large, immigration can only be balanced by the third-order term (crowding) that dominates at high x; this is the only equilibrium, and it is stable. By decreasing k_m, we can drive the system to a regime where two additional equilibrium points appear to the left of the original one. Of these, the left-most (smallest) is also stable, but the system will remain at the original equilibrium point, because its stability has not changed. Ultimately, however, the original point will meet the unstable point and both will disappear, whereupon the system will spontaneously transition to the remaining equilibrium point at the comparatively small value of x. If we now reverse the direction of change in k_m and drive the system back to the regime with three equilibria in total and two stable ones, it will remain in the left-most equilibrium until this equilibrium disappears by merging with the unstable one. At that point, the system will spontaneously transition to the original, high-biomass steady state. Overall, we observe that the state of the system can depend on the history of the variation in its parameters, and there is a domain in the parameter space where the biomass can follow different trajectories depending on the direction in which the control parameter is varied. Such behavior is termed hysteresis. The classical univariate population study with important hysteresis effects is the budworm outbreak model (Ludwig et al. 1978), to which the reader interested in a more concrete ecological treatment can be referred.

2.2 Growth with Generations: Oscillations and Chaos

Population dynamics characterized by die-off of adult organisms following the breeding season are fairly common across the clades of life and can be presumed to be a shared feature of habitable planets with pronounced seasons. In the simplest case, biomass x_i of adult organisms of generation i depends only on the biomass observed in the immediately preceding generation:

$$x_i = f(x_{i-1}).$$

In this instance, equilibria are represented by the roots of $f = 1$. To determine the stability of these equilibria, we can linearize f in the vicinity of each root x^*:

$$\delta_i = x_i - x^*, \quad \delta_i = \delta_{i-1} \left. \frac{df}{dx} \right|_{x^*}.$$

Since the derivative of f acts on perturbation δ like the factor in a geometric progression through the generations that follow, $|df/dx| < 1$ means that the equilibrium is stable, and $|df/dx| > 1$ means that it is unstable.

Up to this point, the behavior of the models with generations has been analogous to the univariate continuous-time models considered in the last section. However, even the simple model with non-overlapping generations can have qualitatively different regimes in addition to the ones already described. Although the deviation from the equilibrium will eventually die down in all instances of $|df/dx| < 1$, it does so in a monotonic manner only if the derivative is positive. If the derivative is negative, the sign of the deviation from the equilibrium point alternates with each generation. In other words, we would observe damped oscillations around the stable steady state. The manner of the departure from the point of unstable equilibrium can similarly be monotonic or in a form of oscillations of increasing amplitude, depending on the sign of df/dx when $|df/dx| > 1$.

Models with generations can have steady states other than equilibrium points. One of the classical models (May and Oster 1976) has the form

$$x_i = x_{i-1} + k_g x_{i-1} - k_d x_{i-1}^2,$$

where $k_g > 0$ is a growth coefficient and $k_d > 0$ s a die-off coefficient. This model, while it is a finite difference version of the polynomial differential equation models from the last section, reflects a slightly different biology than any of the preceding examples. It is suited to describe reproduction that is not limited by individuals' encounters (positive linear term) but at the same time displays a negative effect of crowding (negative quadratic term). Expressing x in the units of the non-zero equilibrium biomass,

$$\bar{x} = \frac{x}{x^*}, \quad x^* = \frac{k_g}{k_d},$$

we can see that the dynamics is controlled by a single parameter, k_g:

$$\bar{x}_i = f(\bar{x}_{i-1}) = \bar{x}_{i-1}\left(1 + k_g\left(1 - \bar{x}_{i-1}\right)\right)$$

We can further note that $df / d\overline{x}$ at $\overline{x} = 1$ (equilibrium) equals $1-k_g$. Thus, the equilibrium is stable for $0 < k_g < 2$ and unstable when $k_g > 2$. To consider the dynamic possibilities afforded by the discrete generations, let us plot f alongside $f^{(2)}$, which is f compounded with itself (*i.e.*, applied sequentially twice). As can be seen in Fig. 2.2—where the bisector indicating the equality of the function value with the argument (*i.e.*, the equality of the reduced biomass $\overline{x}$ one or two generations apart) is also plotted for reference—when k_g approaches 2 from below, $f^{(2)}$ intersects the bisector in the same point as f and its derivative is less than 1. This reflects the fact that biomass size $\overline{x} = 1$ will be stably repeated every two generations as long as it is stably repeated every generation. As k_g crosses above 2, however, the shapes of both f and $f^{(2)}$ indicate a loss of stability at $\overline{x} = 1$. At the same time, two new points arise where $f^{(2)}$ intersects $\overline{x}_i = \overline{x}_{i-2}$. (It is said that a bifurcation takes place.) Each of these points is therefore repeated every second generation. In other words, a two-point cycle is born—a new type of dynamics that is not afforded by univariate models in continuous time. Such a stable cycle, which can attract the trajectory of the system, is referred to as a limit cycle. The slope of $f^{(2)}$ at the two cycle points (Fig. 2.2) is below 1 while being positive, and so demonstrates that each of these points, and therefore the cycle that is comprised of them, is stable.

Continuing to increase k_g in this model, we observe that $df^{(2)}/dx$ at the period-2 stable points decreases until it becomes smaller than -1 (Fig. 2.3). As these points become unstable at period 2, so do they also as points of a degenerate period-4 cycle, as evidenced by the plot of $f^{(4)}$. At the same time, four distinct period-4 stable points are born (Fig. 2.3). Plotting the time course of a simulation starting at some arbitrary biomass size (Fig. 2.4a), we confirm that after a brief initial period, the model trajectory settles into a four-point cycle in which the biomass oscillates about the original "period-1" equilibrium ($\overline{x} = 1$), taking values that alternate above and below it with every new generation. Analysis shows (May and Oster 1976) that

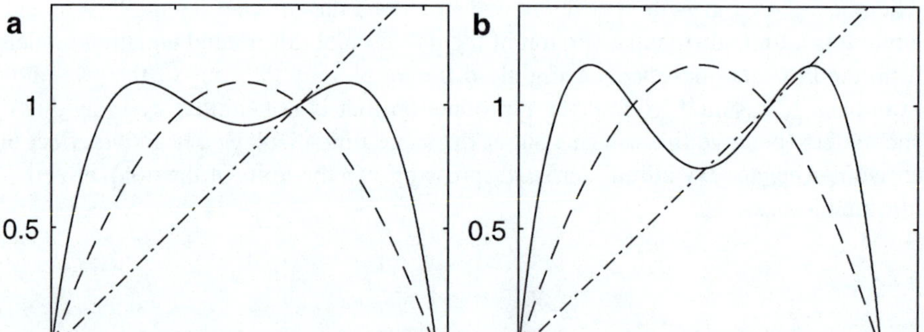

Fig. 2.2 Reproduction function f (*dashed line*) and its second iteration $f^{(2)}$ (*solid line*). The bisector where the function is equal to the argument is plotted for reference (*dash-dot line*). (**a**) $k_g = 1.8$, (**b**) $k_g = 2.2$

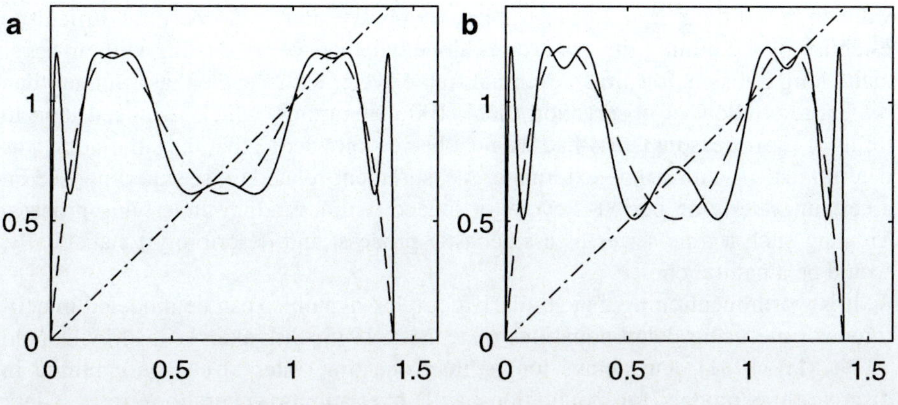

Fig. 2.3 Second-iteration ($f^{(2)}$, *dashed line*) and fourth-iteration ($f^{(4)}$, *solid line*) reproduction function. *Dash-dot line*, bisector for reference. **(a)** $k_g = 2.4$, **(b)** $k_g = 2.53$

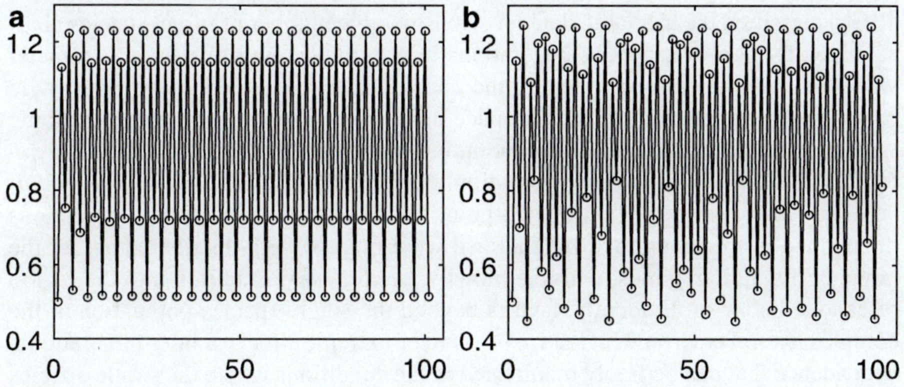

Fig. 2.4 Time course of the model with generations, starting with $\bar{x} = 0.5$. **(a)** $k_g = 2.53$, **(b)** $k_g = 2.6$

period-doubling of this kind continues with the increasing growth parameter, but the parameter intervals leading to each new doubling progressively shorten.

It can be noted that although the standpoint afforded by our knowledge of the dynamics that generate the time series leads us to speak in this case of a single period-4 cycle, a biologist presented by data that have the appearance of the time series in Fig. 2.4a would be equally justified to consider that the dynamics may be that of a comparatively small-amplitude period-4 cycle superimposed on a larger-amplitude period-2 cycle. From such an independent observer's viewpoint, time courses generated with the subsequent period-doubling might appear multiperiodic insofar as the very presence of a repeated pattern remains recognizable.

According to the definition adopted in the mathematical analysis of this type of models, the regime that is called chaotic is reached by the model considered here at around $k_g = 2.57$. Simulating the model with only a somewhat larger parameter value than last time ($k_g = 2.6$, Fig. 2.4b), we observe that it has become difficult to establish by examining the time series alone whether we are dealing with an especially long transient toward an eventual stable cycle, with a cycle that is longer than the entire window of observation (here, 100 generations, which is indeed long in terms of Earth seasons), or—had we no advance knowledge that the simulation was deterministic—with strong external or measurement-related noise superimposed on a certain underlying period-2 cycle, or indeed with a random underlying process. Treating such a time series as a stochastic process, and describing it statistically, would be a natural choice.

It is worth mentioning that similarly complex dynamics can be modeled in continuous time using delay equations (e.g., May 1975; Stirzaker 1975; MacDonald 1979; May 1981), and delays longer than one time interval can be captured in discrete-time models for populations with overlapping generations (e.g., Clark 1976; Getz and Haight 1989). How applicable, however, are the different regimes (stable equilibrium, oscillations, chaos) of the models with generations to explaining the observable biological dynamics? To answer this question, Hassell et al. (1976) classified the available data on a considerable number of insect species characterized by non-overlapping generations in accordance with the fitted parameter values of a model closely related to the one we have considered above. Most wild populations fell into the parameter-space domain corresponding to the monotonic type of stability of the equilibrium point but, notably, the mosquito *Aedes aegypti* fell into the domain of damped oscillations, while the beetle *Leptinotarsa decemlineata* was found in the region of two-point limit cycles. Experimental populations maintained in the laboratory, by contrast, were more widely distributed across the range of dynamic regimes, with the blowfly *Lucilia cuprina* falling into the region of chaotic behavior. Evidently, factors beyond the single-species population in the complex natural environment tend to prevent the extreme effects of inter-generational dependence that can be freely manifested in the conditions where the single-species approximation is exact. Confirming this view is the observation that the mentioned beetle species (the Colorado potato beetle) is the one whose relationship with its food source in the agricultural setting, while occurring outside the laboratory, can be seen as similarly artificial, or at least evolutionarily recent. Inter-species dynamics will be considered explicitly in a later chapter, once the many mechanisms that shape and govern the unique separate species beyond their simple biomass growth have been outlined.

References

Clark CW (1976) Mathematical bioeconomics. Wiley, New York

Edelstein-Keshet L (1988) Mathematical models in biology. Random House, New York

Getz WM, Haight RG (1989) Population harvesting demographic models of fish, forest, and animal resources. Princeton University Press, Princeton, NJ

Hassell MP, Lawton JH, May RM (1976) Patterns of dynamical behaviour in single-species populations. J Anim Ecol 45:471–486

Hastings A (2000) Population biology. Springer, New York
Kot M (2001) Elements of mathematical ecology. Cambridge University Press, Cambridge
Ludwig D, Jones DD, Holling CS (1978) Qualitative analysis of insect outbreak systems: the spruce budworm and forest. J Anim Ecol 47:315–332
MacDonald N (1979) Time lags in biological models. Springer, New York
May RM (1975) Stability and complexity in model ecosystems. Princeton University Press, Princeton, NJ
May RM (1981) Theoretical ecology. Blackwell, Oxford
May RM, Oster GF (1976) Bifurcations and dynamic complexity in simple ecological models. Am Nat 110:573–599
Murray JD (2002) Mathematical biology. Springer, New York
Riznichenko GY (2011) Lekzii po matematicheskim modelyam v biologii (lectures on mathematical models in biology). RHD, Moskva
Stirzaker D (1975) On a population model. Math Biosci 23:329–336

Chapter 3
Species and Speciation

3.1 Selection-Mutation Equilibria

Central to the development of a cohesive and encompassing theory of living matter is the concept of species as diverse and defined forms that the living matter can take. Indeed, every subsequent approach to be laid out in our survey of quantitative general biology relies on existence of a mechanistic structure and a set of biophysical parameters that are representative of some identifiable quantity of living matter. Even the simple biomass growth models considered in the last chapter implied that matter with specific kinetic properties could be singled out for analysis, and that is possible only when the living matter is subdivided into at least tolerably well-defined varieties.

The views on the nature and origin of species have historically swung away from, and then back towards (Mallet 2005) the ones introduced by Darwin and Wallace. Mathematical representation played an important role in the analysis and evaluation of the relative plausibility of concepts related to the place of species in the evolutionary theory (Servedio et al. 2014). Here we will focus on a circle of notions—dealing with mutation, selection, and reproductive isolation—that arose from the beginning in a precise mathematical form and from the beginning were focused on the definition and genesis of species. Although recent, they can be seen as genetically informed formalizations of Darwin and Wallace's, or, in any event, capable of quantitatively explaining the most elementary of the originally envisioned speciation processes: the continuous rise of varieties and the extinction of intermediate forms.

Although the terminology is still not entirely settled (owing to the diversity of ways that the concept of species is utilized in biological research), the least prescriptive accepted definition of a species is as a genetic cluster (Shapiro et al. 2016), *i.e.*, as a type of individuals that have genotypes belonging to a statistically distinct group. This definition presupposes not only variation within each species, but also plurality of species. The quantitative models to be considered, while abstract and

I. Maly, *Quantitative Elements of General Biology*,
https://doi.org/10.1007/978-3-030-79146-9_3

simple in their construction, are capable of recapitulating the emergence of just such population-genetic patterns and therefore appear to capture the most fundamental aspects of speciation.

First, we turn to the question of the genetic distribution that constitutes a species maintained by adaptive selection. Its form and properties in a number of fundamental cases have been laid out systematically by Higgs (1994). The case of a single haploid locus demonstrates the basics of mutation-selection equilibrium and distribution in the genetic space. In this model, the high-fitness allele *A* is characterized by its frequency α in the population, and the alternative allele *a* by frequency $\beta = 1 - \alpha$ (*a* can stand for any of the alternative, lower-fitness alleles, in which case β stands for the frequency of all such alternative alleles combined). Organisms with allele *A* reproduce at a rate $w_0 = 1$ per generation, and organisms carrying any alternative allele reproduce at a rate $w_1 = 1 - s$, where $s > 0$ is the selection coefficient. Allele *A* mutates into allele *a* at a rate *u* per generation, and we are neglecting back mutations, which are unlikely if the locus is not a single nucleotide but itself a long genomic sequence. This leads to the following equation for iteration between generations *t*:

$$\alpha_{t+1} = \frac{(1-u)\alpha_t}{\alpha_t + (1-s)\beta_t}$$

As long as $u \leq s$, α_t converges to $1 - u/s$. Correspondingly, β_t in this instance converges to u/s. The pair of allele frequencies $(1 - u/s, u/s)$ is a simple example of a stationary distribution in the genetic space that demonstrates a defined degree of variability that evolves within the species when the balance of mutation and selection is reached. Note, however, that if *u* exceeds *s*, then α_t converges to 0, *i.e.*, the high-fitness allele will be lost from the population. The condition $u = s$ is said to represent an error threshold in the reproduction of organisms governed by selection and mutation.

The error-threshold behavior of a multi-locus haploid genome is a useful stepping stone to more complex biological systems. In this case, there may be *k* loci with non-optimal alleles, out of the total of *L*. The fitness of an organism with such a genotype can be denoted w_k. As before, and without any loss of generality, we will take w_0 to be 1, and consider the frequencies of the possible genotypes. After denoting the combined frequency of all genotypes with *k* non-optimal alleles C_k, the dynamic equation for iterating over generations of the haploid organisms is

$$C_j(t+1) = \frac{1}{W}\sum_k M_{jk} w_k C_k(t), \tag{3.1}$$

where *W* is the mean fitness of the population and *M* is the mutation matrix. Again neglecting back mutations,

$$M_{jk} = \binom{L-k}{j-k} u^{j-k}(1-u)^{L-j}, \quad k \leq j \leq L. \tag{3.2}$$

If, however, L is large and u is small, then M has the form

$$M_{jk} = e^{-U} \frac{U^{j-k}}{(j-k)!}$$

whether the back mutations are included or not. Here we have introduced $U = uL$, the total genomic mutation rate. Equation (3.1) can be iterated numerically for any given initial condition $\{C_k(1)\}$ and fitness landscape $\{w_k\}$.

A simple fitness landscape characterized by one "master" genetic sequence (Swetina and Schuster 1982) has an isolated peak whose height is w_0, and any other sequence of alleles is assumed to have fitness $1 - s$, as with the alternative alleles in the single-locus model just analyzed. In the large-L approximation, the genetic distribution converges in this instance to

$$C_j = \sum_{k=0}^{j} \frac{U^{j-k}}{(j-k)!} w_k C_k.$$

If s and U are both small, the stationary distribution approaches the power law $C_j = C_0\,(U/s)^j$. Normalization of the frequencies then requires that $C_0 = 1 - U/s$ if $U < s$. This is an extension of the single-locus result, with the frequency of the master sequence, C_0, having the same form when expressed in terms of the genomic mutation rate U as the favored allele A had in terms of the single-locus mutation rate u. In the master-sequence model, the power law approximation suggests that the error threshold is reached when $U = s$, which is again a genomic analog of the single-locus case. However, the master genetic sequence does not fully disappear from the population in the exact solution.

Although the master-sequence landscape may appear simplistic as a model for natural selection, the solution obtained in this case demonstrates how a species maintained by selection of even a singular favored genotype will exist as a cloud of mutants with a shape, in the genetic space, that is precisely defined by the selection's strength relative to mutation. When the master-sequence model was originally developed in application to molecular evolution, such a stationary distribution was termed a quasi-species (Eigen et al. 1988), a concept soon afterwards adopted in studies of viral evolution. Advances in genome sequencing have since made it clear that all biological species may be "quasi-species" when judged against any species paradigm that emphasizes genetic uniformity.

The next approximation for a realistic fitness landscape that can be considered is multiplicative. In it, the reduced fitness $1 - s$ is applied as a factor for each non-optimal allele present in the genetic sequence, so that $w_k = (1 - s)^k$. In this case, the time-independent solution to Eq. (3.1) is

$$C_j = \binom{L}{j} (u/s)^j (1-u/s)^{L-j}.$$

At each locus, we have the optimal allele at frequency $\alpha = 1 - u/s$ and the mutant allele at frequency $\beta = u/s$ like in the single-locus model, a property that is by no means general and was not observed in the master-sequence landscape. The optimal-sequence concentration is $C_0 = (1 - u/s)^L$, so while there is an error threshold at $u = s$ as before, the transition is less pronounced for large genomes with multiplicative fitness. In fact, when L is large, we can operate with the genome-wide mutation rate U and write

$$C_j = e^{-U/s} \frac{(U/s)^j}{j!},$$

which describes a population that is smoothly losing its concentration around the optimal sequence with increasing mutation rate relative to selection. These reference cases having been established, we can proceed to the discussion of diploid populations.

To study diploid genetic distributions established under the action of selection and mutation, we will consider models where each locus can contain two highest-fitness alleles (*AA*), two reduced-fitness alleles (*aa*), or one of each kind (*Aa*). Note that while this representation may resemble some biological descriptions where the two alleles are specified on the genetic level, it actually deals with *types* (*A* or *a*) of alleles whose fitness but not necessarily genetic identity is prescribed. So, as in the single-locus model above, there may be many genetically different mutations creating *a* from *A*, but only their combined rate and identical fitness impact are considered in the model. Keeping this in mind, we will call the *AA* and *aa* loci homozygous and the *Aa* loci heterozygous.

In the diploid case, instead of Eq. (3.1), we have

$$C_{jk}(t+1) = \frac{1}{W} \sum_{n=0}^{L} \sum_{m=0}^{L-n} M_{jknm} w_{nm} C_{nm}(t),$$

where index jk refers to genotypes with j homozygous mutant loci and k heterozygous loci. The mutation matrix takes the form

$$M_{jknm} = \binom{m}{j-n} \binom{L-n-m}{j-n+k-m} u^{j-n} (1-u)^{m-j+n} (2u)^{j-n+k-m} (1-2u)^{L-j-k}$$

for mutation from genotype nm into jk, assuming that the likelihood of mutation $AA \rightarrow aa$ in one generation in negligible. The multiplicative fitness in the diploid case is modeled as $w_{jk} = (1 - s)^j (1 - hs)^k$, with dominance coefficient h describing the degree of selection against a heterozygous locus compared with a homozygous mutant locus.

The equilibrium distribution of genotype frequencies takes different forms depending on the parameter values. It can be approached by observing that if the three types of loci—*AA*, *Aa*, and *aa*—are found at finite frequencies *x*, *y*, and *z*, the equilibrium distribution can be written as

$$C_{jk} = \frac{L!}{j!k!(L-j-k)!} x^{L-j-k} y^k z^j, \tag{3.3}$$

which leads to the steady-state values

$$x = \frac{(h - u/s)(1 - 2u/s)}{h + (1-2h)u/s}, \quad y = \frac{2u/s(1-2u/s)}{h+(1-2h)u/s}, \quad z = \frac{2(u/s)^2}{h+(1-2h)u/s}$$

in the limit of small *u* and *s*. In this situation, all possible genome types *jk* are represented, and, as in the haploid case with the multiplicative landscape, the steady-state fractions of the loci types are the same whether we are considering an isolated locus (by putting $L = 1$) or the whole genome. The positivity of *x*, *y*, and *z* requires that u/s be smaller than ½ as well as smaller than *h*. When u/s reaches the smaller of ½ and *h*, homozygous loci with favored alleles disappear ($x = 0$). If $h \geq ½$, *i.e.*, if the fitness of *Aa* is closer to that of *AA*, this error threshold is reached at $u/s = ½$. Then we have $z = 1$, and the only remaining genotype has *aa* at all loci. If, however, $h < ½$ (the fitness of *Aa* is closer to that of *aa*), the threshold where Eq. (3.3) breaks down is reached already at $u/s = h$. In this instance, when *x* becomes 0, both *y* and *z* are finite and we should write

$$C_{j,\,L-j} = \frac{L!}{j!(L-j)!} y^{L-j} z^j,$$

which leads to $y = 1 - z$, $z = u/s(1 - h)$. In this regime, the genomes consist of combinations of *Aa* and *aa* loci, and the fittest organisms are purely heterozygous. It may be emphasized that although biological situations are possible whereby heterozygotes are intrinsically selected for, this is not the nature of the fitness landscape considered here, and the last result is obtained because heterozygotes emerge as the fittest individuals when the favored homozygous loci cannot be maintained. The latter regime holds while $h < u/s < 1 - h$ (and $h < ½$). At $u/s = 1 - h$, *z* becomes 1, and we again have only *a* alleles. Thus, with the weakening of selection relative to mutation, the diploid passes through one or two thresholds depending on the allele dominance, leading to a complete loss of the favored alleles either directly or through a state where only the homozygous high-fitness loci are lost.

The above treatment was developed for diploids propagating by direct replication. The prevalent species concepts, by comparison, are oriented toward sexual populations, appropriately for much of advanced life. In closing of this section we will therefore consider an extension of the developed analysis to genetic distributions arising under selection and mutation on the background of random mating, *i.e.*, within a reproductively defined species. More precisely, we will consider a model with random fusion of the gametes from the last generation and with the production of gametes being linked to the fitness of the diploid organism, still following Higgs (1994).

Taking into account the fact that the numbers of aa and Aa loci arising from fusion of gametes with m and n lower-fitness alleles are related as $2j + k = m + n$, the probability of a diploid of type jk arising from such an event can be written as

$$P_{jk}^{nm} = \binom{n}{j}\binom{L-n}{m-j}/\binom{L}{m}.$$

If so, and such gametes exist at frequencies X_m and X_n, their fusion gives rise to the diploid frequencies in the next generation according to

$$C_{jk} = \sum_n \sum_m X_n X_m P_{jk}^{nm}.$$

When no mutation occurs, a type-jk organism produces gametes with m alleles a with probability

$$g_m^{jk} = \binom{k}{m-j}\frac{1}{2^k}, \qquad j \le m \le j+k.$$

Not stipulating that mutation occurs in the gametes, it can nonetheless be accounted for on a per-generation basis using the already introduced haploid mutation matrix M (Eq. 3.2), and the production of gametes described as

$$X_n = \frac{1}{W}\sum_m M_{nm} \sum_j \sum_k C_{jk} w_{jk} g_m^{jk}.$$

A stationary solution can be found that, like the other multiplicative-fitness models, exhibits no linkage disequilibrium. The allele-type frequencies being denoted by α for A and β for a as before, the stationary gamete frequencies have the form

$$X_n = \binom{L}{n}\alpha^{L-n}\beta^n,$$

and the diploid-stage genotype frequencies follow the form observed in the asexual diploid (Eq. 3.3) with Hardy-Weinberg locus type frequencies $x = \alpha^2$, $y = 2\alpha\beta$, $z = \beta^2$. The condition for such a stationary state in the sexual model, again assuming small s and u, is that

$$\beta = \frac{-h + \sqrt{h^2 + 4(1-2h)u/s}}{2(1-2h)}, \qquad h \ne \tfrac{1}{2},$$

and $\beta = 2u/s$ when $h = ½$. We can see that alleles A are lost when u/s reaches $1 - h$. This happens through continuous reduction if $h < \frac{2}{3}$. When $h > \frac{2}{3}$, the above solution for β becomes complex as u/s reaches $h^2/4(2h - 1)$, but the trivial stationary solution with $\beta = 1$ always exists and the system will transition to it in a discontinuous fashion. Thus, in the sexual diploid case the regime with only the Aa and aa loci types does not exist, and there is a single error threshold leading to all-mutant genotypes directly from the state where all types are represented. The shape of the distribution in the jk space is illustrated in Fig. 3.1.

Altogether, the models considered in this section demonstrate that well-defined and, generally, broad genetic distributions are established within a reproductively

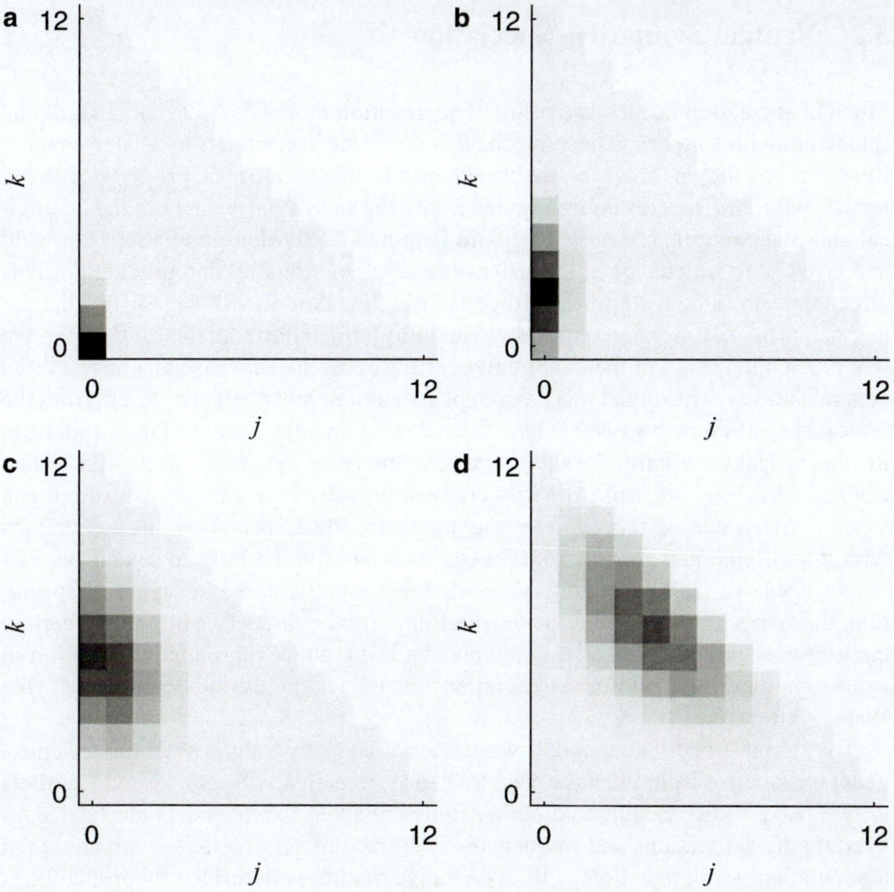

Fig. 3.1 Equilibrium distributions of diploid genotypes in a randomly mating population. The density of genotypes in the jk space is shown in grayscale with the highest values represented by black. In all examples, $L = 12$ and $h = 0.25$. (**a**) $u/s = 0.005$, (**b**) $u/s = 0.03$, (**c**) $u/s = 0.1$, (**d**) $u/s = 0.3$. (After Higgs 1994)

cohesive population (species) under the action of selection and mutation. These distributions, in turn, define the statistical distribution of any genetically determined property of organisms in the population, such as those metabolic, kinetic, mechanical, and behavioral properties that underlie the physiological and ecological dynamics to be considered in the later chapters. We have observed that the steady-state frequency of the selectively favored alleles can exhibit a strongly nonlinear response to the relative power of selection and mutation—even when each of these processes has a weak impact on the generational time scale. At the same time, presupposing the reproductive cohesion excludes formation of new species in models of this kind. Mechanisms leading to emergence of reproductive isolation and species formation will be considered next.

3.2 Neutral Sympatric Speciation

Much of speciation occurs as a result of segregation into ecological and geographic (planetographic) niches. These mechanisms will be considered in a later chapter. Research has shown, however, that speciation is also an intrinsic property of living matter, with new species being capable of arising in sympatry (in both the ecological and planetographic sense). It should be noted firstly that populations can split and give rise to isolated genetic clusters as a result of clonality in reproduction. This effect appears to be a significant force behind the taxonomy of organisms that are asexual, primarily asexual, or otherwise exhibit limited horizontal gene transfer. We will touch upon some of the quantitative results in the asexual case that have a more general validity. The clonal mechanism of speciation, however, is different from the typical speciation pathways that have been defined in advanced life forms and, from the theoretical viewpoint, does not represent emergent dynamics. In the rest of this chapter, therefore, we will primarily concern ourselves with mechanisms that can sustain emergence of reproductive isolation (and thus speciation) in populations with a high potential for horizontal gene transfer—particularly in sexual populations capable of random mating. Vis-à-vis the allopatric and other types of speciation, these mechanisms have fundamental importance. Indeed, while under certain circumstances to be delineated here they can bring about reproductive isolation in sympatry, they also reinforce speciation occurring via the other pathways (the Wallace effect).

The simplest one-locus models with assortative mating and a potential for emergent reproductive isolation have been treated systematically by Boake and Gavrilets (1998). Like those encountered earlier in this chapter, these models consider non-overlapping generations and frequencies x, y, and z of genotypes AA, Aa, and aa in a population with mean fitness W. Assortative mating is described by probabilities p_{ij} of successful mating in a pair of individuals representing genotypes i and j (numbered in the above order), yielding the following equations for Mendelian iteration over generations:

$$Wx_{t+1} = p_{11}x_t^2 + p_{12}x_t y_t + \frac{1}{4}p_{22}y_t^2,$$

$$Wy_{t+1} = p_{12}x_t y_t + 2p_{13}x_t z_t + \frac{1}{2}p_{22}y_t^2 + p_{23}y_t z_t, \tag{3.4}$$

$$Wz_{t+1} = \frac{1}{4}p_{22}y_t^2 + p_{23}y_t z_t + p_{33}z_t^2.$$

Note that we are considering only situations where representatives of each sex are produced and survive in equal numbers.

The steady-state solutions to Eq. (3.4) can be categorized into classes that differ in the compactness of the genotypic distribution and reproductive cohesiveness of the populations that they represent. The first regime describes polymorphic (x, y, $z > 0$) populations where all genotypes can interbreed either directly or via the intermediate genotype. The attractors of Eq. (3.4) in this case may take the form of symmetric ($x = z$) or asymmetric ($x \neq z$) equilibria and period-2 cycles, and describe reproductively cohesive, single species. The second regime is convergence to a monomorphic population: either the one with $x > 0$, $y = z = 0$, or the mirror case, where $x = y = 0$ and $z > 0$. Disappearance of genotypes as a result of such convergence is a simple example of formation of a tight genetic cluster, but the population remains a reproductively cohesive, single species. The monomorphic state where the population consists of only genotype *AA* will be stable with respect to any arising alleles *a* if $p_{11} > p_{12}$. A mirror condition exists for local stability of the other possible monomorphic equilibrium, where essentially all individuals are *aa*: $p_{33} > p_{23}$. In other words, for the rare allele not be become established in the population, the prevalent homozygotes must mate with their own type with a likelihood that is higher than that of successful matings involving the rare heterozygotes.

However—and this represents the third regime, one that is the most relevant to our subject—both monomorphic equilibria can be stable at the same time. This only requires that $p_{11} > p_{12}$ and $p_{33} > p_{23}$ simultaneously, and when this is the case, the analysis suggests that no other equilibria should be stable. If additionally $p_{13} = 0$, Eq. (3.4) describes a system with two morphs that do not interbreed directly and can exist as alternative species occupying the same niche. Moreover, one can see that along with the bistability and a potential for externally driven or stochastic switching that it implies, both *AA* and *aa* genotypes (species, in this case) can stably coexist without mixing once the heterozygotes (hybrids) disappear. Individual-based stochastic simulations (Boake and Gavrilets 1998) demonstrate that these scenarios of transitioning to a new species (from *AA* to *aa*, for example) and from one to two sympatrically isolated species are indeed realizable when the total population size is reduced, such as in a previously monomorphic, reproductively cohesive species that is passing through a population bottleneck in presence of a perturbation (mutation) causing the alternative allele to appear. At the same time, the realizability of the regime with the simultaneously stable monomorphic equilibria is borne out by measurements of the mating probabilities p_{ij} in insect species. Summing up, sympatric speciation is supported even by assortative mating mechanisms that can be controlled by a single locus.

A systematic analysis of multi-locus systems with reproductive isolation has been carried out by Higgs and Derrida (1992). To consider these fundamental dynamics, we can assign numerical designations $S_i = 1$ and $S_i = -1$ to each pair i of alternative alleles. The genome will be haploid, but, as pointed out by Gavrilets (1999), a model formulated for haploids mating assortatively based on their genetic distance applies equally to randomly mating diploids with post-mating isolation that selects against heterozygosity. As before, in this treatment there are L loci, and mutation occurs at rate u per locus per generation. The alleles mutate from 1 to -1 and vice versa independently over the genome, while the number of individuals is retained at N from generation to generation. The model focuses on the measure of similarity between individuals that can be expressed as

$$q^{\varphi\psi} = \frac{1}{L}\sum_{i=1}^{L} S_i^{\varphi} S_i^{\psi}$$

for the pair of individuals φ and ψ.

Although the approach is intended for analysis of mating, it is helpful to consider first what happens with his formalism in the asexual case. Then, the model is iterated by choosing a random organism that becomes the parent of one offspring, until N offspring constituting the next generation have been produced. In this way, each extant organism has only one parent, but the distribution of the number of offspring that this organism may itself produce is Poisson with the mean 1. The expectation of the allele at locus i in the offspring φ is

$$E\left(S_i^{\varphi}\right) = \frac{1}{2}\left(1 + e^{-2u}\right) S_i^{P(\varphi)} - \frac{1}{2}\left(1 - e^{-2u}\right) S_i^{P(\varphi)} = e^{-2u} S_i^{P(\varphi)},$$

where $P(\varphi)$ denotes the parent organism. Transitioning from this to the expected similarity between individuals φ and ψ in the next generation, we can note that owing to the independence of mutation between the L loci, the variance around the expectation will vanish in the limit of large L. As a result, we can simply write

$$q^{\varphi\psi} = e^{-4u} q^{P(\varphi)P(\psi)}.$$

It follows that insofar as we are interested in the evolution of similarity in the population with large L, the updating rule just derived allows operating with N-by-N matrices of elements q—similarity matrices—instead of the complete individual genomes.

Numerical simulations show that the distribution of $q^{\varphi\psi}$ values in the asexual case consists of well-separated, sharp peaks that drift from one to zero. Naturally, these peaks correspond to branches and stems of the clonal tree that could be traced backward in time from the extant individuals. In this way, the shape of the distribution illuminates the taxonomy of the organisms in question. While the specific form of the distribution at the given time may be unpredictable, the peaks in it are not emergent but a direct consequence of the clonal reproduction. The mean similarity established in the asexual case, however, turns out to be an important parameter of more general validity. To derive it, we can note that if the probability $1/N$ that two individuals have the same parent is small, the time-averaged population mean of the genetic similarity must satisfy the equation

$$q_0 = e^{-4u}\left[\frac{1}{N} + \left(1 - \frac{1}{N}\right) q_0\right],$$

which works out to $q_0 = 1/(1 + 4uN)$ when u is also small.

Returning to the sexual case, the model can be iterated by choosing two parents at random. The next-generation individual is created that inherits with equal probability the allele of one or the other parent independently for each locus, either identically or with mutation as defined earlier. Noting that there are four equally probable combinations of parents behind each pair of loci in the individuals which we are comparing, the above result for iterating the off-diagonal elements of the similarity matrix can be extended by writing

$$q^{\varphi\psi} = \frac{e^{-4u}}{4}\left(q^{P_1(\varphi)P_1(\psi)} + q^{P_2(\varphi)P_1(\psi)} + q^{P_1(\varphi)P_2(\psi)} + q^{P_2(\varphi)P_2(\psi)}\right),$$

where the lower indices number the parents of each organism.

Even though the population drifts neutrally in the genomic space, the distribution of q in the random-mating model is, as expected, stationary and consists of a single, vastly dominant peak and a few small ones that represent slightly higher genetic similarity. In the main, this recapitulates the behavior analyzed in the last section, and our purpose here is to characterize the overlap distribution corresponding to random mating quantitatively as a reference point before treating the assortative variant of the model. The mean of the overlap distribution with random mating can be found analogously to the approach adopted for asexual populations, by neglecting the $O(1/N^2)$ probability that the parents of two given individuals are the same and noting that the probability that they share one parent is $4/N$:

$$q_0 = e^{-4u}\left[\frac{4}{N}\left(\frac{3q_0 + 1}{4}\right) + \left(1 - \frac{4}{N}\right) q_0\right].$$

The solution for small u turns out to be the same as derived above: the static mean genetic overlap in a randomly mating population is, in the chosen approximation, identical to the one averaged over time in clonal organisms, despite the distribution looking entirely different. The small peaks to the right (higher similarity) of the peak around q_0 represent subpopulations that have recent common ancestors, *e.g.*, at $q = (15q_0 + 1)/16$ for individuals with a common grandparent. Like the main peak, these minor subgroups are stationary in their size and do not drift in the similarity space, in contrast to the clonal picture. Collectively, these results form a background that helps to highlight the effects of assortative mating when it is added to the same formalism.

Assortative mating represents the dynamics where the first parent is chosen, as before, entirely at random, but the second one is then chosen only among the individuals that are genetically similar to the first parent. Specifically, it will be chosen at random among the individuals for which q is greater than a certain cutoff parameter $q_{\min}$. In this way, the selection is neutral with respect to individuals but, depending on the value of $q_{\min}$ and the actual genotype distribution in the population, can be strongly non-neutral with respect to potential mating pairs. Given the properties that have been established in the random mating model, it is clear that when $q_{\min} < q_0$, all individuals extant at steady state mate randomly: the introduced cutoff has no effect, and the steady-state population will remain homogeneous. If, by contrast, $q_{\min} > q_0$, then the homogeneous equilibrium cannot be reached, and simulations show that despite the genetic reshuffling brought about by the sexual mode of reproduction, we will again witness a dynamic picture reminiscent of the clonal case: the distribution of q will display a series of drifting peaks as a steady-state behavior (Fig. 3.2). In other words, the sexual population will break apart into taxonomical sub-families. This behavior represents spontaneous emergence of species and their genealogical relations. Numerical analysis of the assortative mating model shows that there will be more simultaneously existing species when $q_{\min}$ is higher. This is unsurprising, because tighter reproductive isolation means smaller genome-space domains for each species, allowing a greater number of them to co-exist in the population niche of a given size.

Looking closely at the example in Fig. 3.2, we can see that this rather simple simulation output with low $q_{\min}$ (0.65) contains two species peaks above the cutoff, which fluctuate in exact position and size, and a single peak below the cutoff, which represents the genetic overlap between the species. The latter peak drifts predictably toward zero as the independent mutations within the reproductively isolated species accumulate. The dynamic details of the fluctuations exhibited by the species peaks, at the same time, are illuminating of the process of species persistence through generations and its alternatives, extinction and (neutral) evolutionary radiation.

First we may note that each self-organized species (say, species Φ) has its own, internal mean overlap value. At equilibrium, it would be related to the species' size, m_Φ, through a relationship whose form follows that derived for the randomly mating population: $q_0(m_\Phi) = 1/(1 + 4um_\Phi)$. The species would retain its reproductive cohesion as long as m_Φ is sufficiently large, so that $q_0(m_\Phi) > q_{\min}$. However, with its assumed low rate, mutation is not acting rapidly enough to equilibrate the mean genetic overlap in accordance with the species size as the latter fluctuates from generation to generation. Hence the largely uncoordinated fluctuation of the size and position of the species' peaks that was evident in the plotted example. As reviewed by Higgs and Derrida (1992), this is similar to the known behavior of homozygosity and the inbreeding coefficient, which may be quite random except after a stringent population bottleneck.

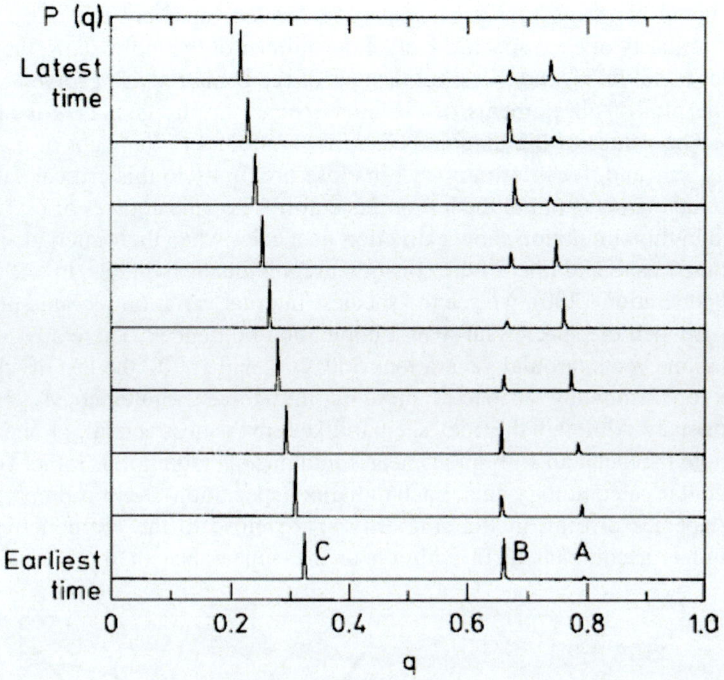

Fig. 3.2 Distribution of inter-individual genetic similarity q in a simulation of assortative mating. $N = 2000$, $q_0 = 0.5$, $q_{\min} = 0.65$. (Reproduced from Higgs and Derrida (1992). Reprinted by permission from Springer Nature Customer Service Centre GmbH: Springer Nature, Journal of Molecular Evolution (https://www.springer.com/journal/239), Genetic distance and species formation in evolving populations, Paul G. Higgs et al., copyright 1992)

The dynamics of the overlap distribution in the population with assortative mating can be further analyzed by reducing it to intra-species and inter-species overlap values. To do this, all individuals must be unambiguously classified into non-overlapping species. This goal is easily achievable in the presence of the reproductive cutoff: one only needs to start with a random individual, find all individuals having an overlap greater than $q_{\min}$ with it, then all individuals having an overlap greater than $q_{\min}$ with the ones already selected. Once there are no more that could be selected, the first species can be considered as fully constituted, and any previously unselected individual can be chosen to start the process anew for the next species. The sorting terminates when there are no more unassigned individuals. If K species have been defined in the population in accordance with the above procedure, then, by analogy with the inter-individual similarity matrix of elements $q^{\varphi\psi}$, a K-by-K matrix of inter-species similarity can be introduced, with elements

$$Q^{\Phi\Psi} = \frac{1}{m_\Phi m_\Psi} \sum_{\varphi \in \Phi} \sum_{\psi \in \Psi} q^{\varphi\psi}.$$

The diagonal elements of such a matrix, naturally, represent the mean internal genetic similarity of each species. Unlike the number of the individuals, the number of species is not fixed, and therefore the size of the Q matrix will evolve.

A time plot of all elements of the inter-species similarity matrix is shown in Fig. 3.3. The values of the diagonal elements are found to the right of $q_{\min}$ on the similarity axis and, for brief intervals, in close proximity to this critical value. The off-diagonal elements fall to the left of this cutoff. The time courses of $Q^{\Phi\Psi}$ that can be traced in this simulation show extinction of species when their internal similarity approaches 1 as the number of the constituent individuals dwindles. In one instance (around generation 1200), we see the species' internal variation rebound off values close to 1, when the species survives a population bottleneck. There are three neutral radiation events (around generations 500, 700, and 1000), the last of which is a complex one, showing a species dividing into three simultaneously or quasi-simultaneously. Although this may seem unlikely in the abstract, Higgs and Derrida (1992) note that such an ambiguous near-simultaneous radiation is rather typical in the molecular paleontology data. Each radiation (speciation) event is prepared by an existing species drifting in the similarity space close to the reproductive cutoff ($q_{\min}$), and is accompanied, in addition to the appearance of the new similarity

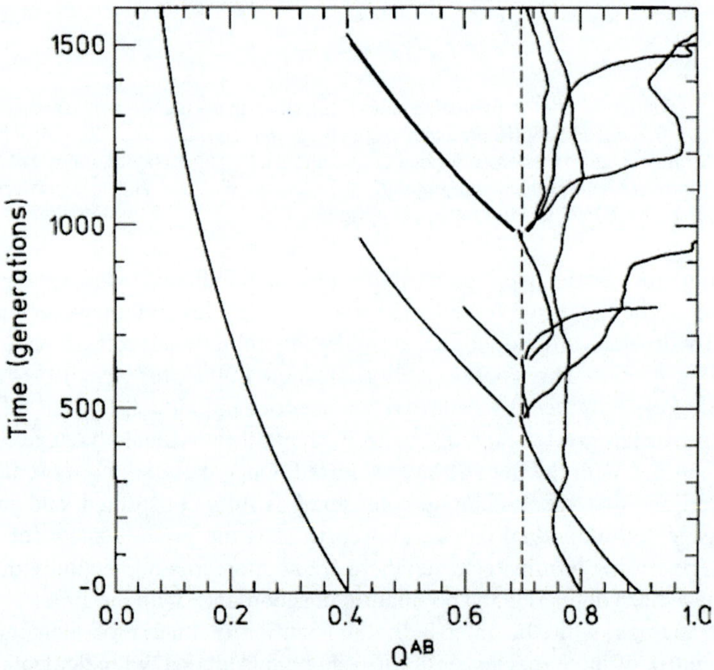

Fig. 3.3 Intra- and inter-species similarity Q time courses in a simulation of assortative mating. $N = 1000$, $q_0 = 0.5$, $q_{\min} = 0.7$. (Reproduced from Higgs and Derrida (1992). Reprinted by permission from Springer Nature Customer Service Centre GmbH: Springer Nature, Journal of Molecular Evolution (https://www.springer.com/journal/239), Genetic distance and species formation in evolving populations, Paul G. Higgs et al., copyright 1992)

courses corresponding to the daughter species, by a start of a new inter-species similarity course to the left of the cutoff value. In the instance of the complex speciation event (near generation 1000), to the left of the cutoff we observe a close family of curves corresponding to the three mean distances between three species of equal age. One species survives the entire length of the simulation without either extinction or radiation, and its similarity with the others is very low at the end (Fig. 3.3). The exact behavior around the speciation events is hard to discern through the lens of the similarity matrix and requires other inspection tools.

A spread function can be introduced that measures the prevalence of genetically mismatched potential mating pairs within a species:

$$\sigma_\Phi = \frac{1}{m_\Phi^2} \sum_{\varphi \in \Phi} \sum_{\psi \in \Psi} \theta\left(q_{\min} - q^{\varphi\psi}\right),$$

where θ is the Heaviside step function. To trace σ_Φ through a speciation event, it is helpful to eliminate the heretofore inconsequential ambiguity in naming species. The individuals being numbered in each generation from 1 to N, we will associate species Φ with individual $\varphi = 1$ and always select $P_1(1) = 1$. While the naming and numbering device has no impact on the actual dynamics, it ensures that the selected species' name (here, Φ) is inherited by one of the daughter species. In the following, we will refer to all species that bear this name over time as one species. Aligning σ_Φ with $\phi_\Phi = m_\Phi/N$ and $Q^{\Phi\Phi}$ over sequential speciation events that occur in the same genealogical branch (Fig. 3.4) reveals that the fluctuations of the species population size take it to high values that result in the genetic overlap dropping briefly below the reproductive cutoff value. Mismatched mating pairs appear, resulting in splitting-off of new species, which abruptly decreases the number of individuals belonging to the species we are tracing. The remaining individuals' mean similarity then increases, always tending to the $q_0(m_\Phi)$ value but never able to catch up with the rapid fluctuations in ϕ_Φ. Speciation continues at irregular intervals, as the fluctuations in the species take it near the boundary of the reproductive cohesion region.

Neglecting the intricate intra-species genetic dynamics revealed by the Higgs-Derrida model, and also the fact that, when introduced by Eldredge and Gould (1972), this concept emphasized allopatry, it can be said that the species' evolution traced by the spread function (Fig. 3.4) exhibits "punctuated equilibria." We will encounter this type of saltatory system dynamics once again in the chapter on the evolution of ecological interactions. The example elaborated here demonstrates that the "punctuated" evolution can also be intrinsic to each species in the quasi-neutral case, when the species' mutual dependence is limited to their genealogical relations and the constraint of the total biomass size.

The above analysis depended on the assumption of the infinite number of loci, which permitted operating with the inter-individual similarity matrix alone. Finite genomes must be traced from generation to generation explicitly. A recent study (de Aguiar 2017) that produced such a complete simulation of the model showed that the overlap distributions with finite genomes are wider when compared with the

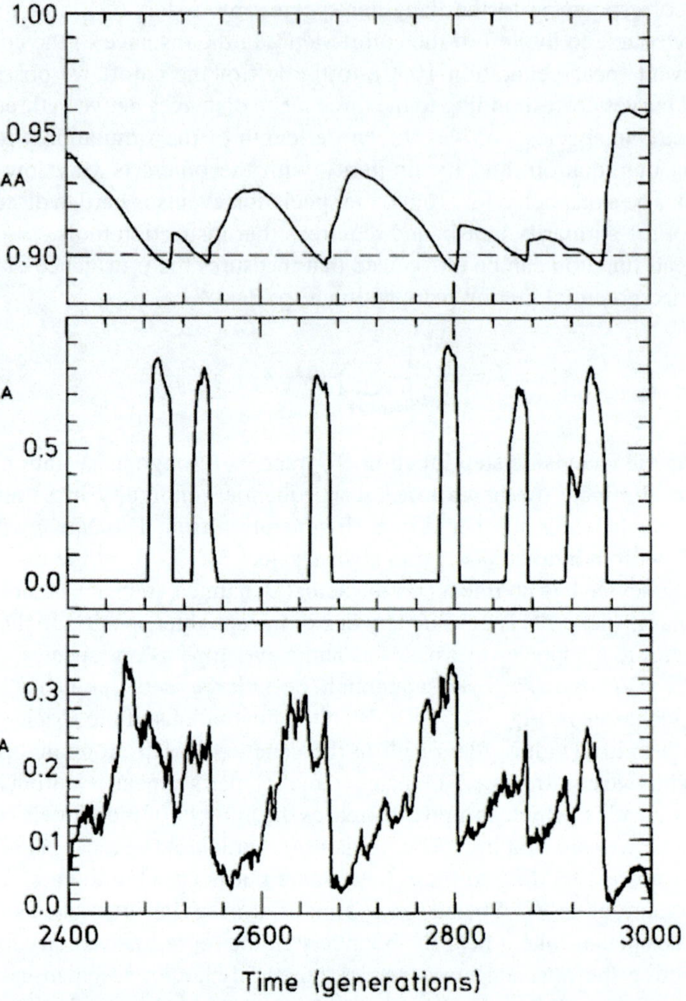

Fig. 3.4 Time courses of intra-species similarity Q, spread function σ, and the species' fraction in the sympatric population, ϕ. $N = 1000$, $q_0 = 0.5$, $q_{\min} = 0.9$. (Reproduced from Higgs and Derrida (1992). Reprinted by permission from Springer Nature Customer Service Centre GmbH: Springer Nature, Journal of Molecular Evolution (https://www.springer.com/journal/239), Genetic distance and species formation in evolving populations, Paul G. Higgs et al., copyright 1992)

infinite case and can appear stalled at $q_{\min}$ instead of displaying speciation when the number of loci controlling the reproductive isolation is too small. The number needed for speciation is related inversely to the mutation rate and directly to the population size. The last result recapitulates the behavior that we observed in the single-locus model applied to emergence of reproductively isolated subpopulations when a species passes through a population bottleneck. Thus, the strong population-size effect appears to be a general feature of the strictly sympatric speciation mechanisms.

References

Boake CRB, Gavrilets S (1998) On the evolution of premating isolation after a founder event. Am Nat 152:706–716

De Aguiar MAM (2017) Speciation in the Derrida–Higgs model with finite genomes and spatial populations. J Phys A 50:85602

Eigen M, McCaskill J, Schuster P (1988) The molecular quasi-species. J Phys Chem 92:6881–6891

Eldredge N, Gould SJ (1972) Punctuated equilibria: an alternative to phyletic gradualism. In: Schopf TJM (ed) Models in paleobiology. Freeman Cooper, San Francisco, CA, pp 82–115

Gavrilets S (1999) A dynamical theory of speciation on holey adaptive landscapes. Am Nat 154:1–22

Higgs PG (1994) Error thresholds and stationary mutant distributions in multi-locus diploid genetics models. Genet Res 63:63–78

Higgs PG, Derrida B (1992) Genetic distance and species formation in evolving populations. J Mol Evol 35:454–465

Mallet J (2005) Speciation in 21st century. Heredity 95:105–109

Servedio MR, Brandvain Y, Dhole S et al (2014) Not just a theory: the utility of mathematical models in evolutionary biology. PLoS Biol 12:e1002017

Shapiro BJ, Leducq JB, Mallet J (2016) What is speciation? PLoS Genet 12:e1005860

Swetina J, Schuster P (1982) Self-replication with errors: a model for polynucleotide replication. Biophys Chem 16:329–345

Chapter 4
Signaling and Control

4.1 All-or-Nothing Response

Morphogenesis, movement, and species interactions to be considered in the rest of this book all depend on the organisms' ability to maintain their internal homeostasis and respond to stimuli arriving from the environment. In fact, although the term "homeostasis" may suggest a state of stability in isolation, in biology such stasis is but a finely controlled, continuous response to the changing environment. Thus, signal processing and control processes on the organism or cell level represent functions that are intertwined, often share the same molecular substrate, and rely on closely related types of nonlinear dynamics. Perhaps the most elementary of these is the type of signaling that occurs on the cell level is known as "all-or-nothing response." It represents an extreme case of nonlinear response of the cell's homeostasis (cytophysiological state) to a signal perceived in the cell's environment, which in the case of a multicellular organism may be a physiological signal internal to the organism.

One general case of strongly nonlinear response is associated with chains of enzymatic reactions, which abound among the metabolic and signaling components of the cell's biochemistry. Huang and Ferrell (1996) highlighted the kinetic features leading to emergence of such a response in sequences of protein phosphorylation reactions known as phosphorylation cascades. One conserved phosphorylation cascade involves—with variations concerning the specific proteins performing each function—a mitogen activated protein kinase (MAPK, *e.g.*, Erk2), a MAPK kinase (MAPKK, *e.g.*, Mek1), and a MAPKK kinase (MAPKKK, *e.g.*, Mos). Such a cascade, wherein MAPKKK phosphorylates MAPKK, activating it and making it capable of phosphorylating MAPK, can transmit a signal from a plasma membrane receptor to the nucleus that triggers a transition of the cell to mitosis, but is also found to participate in other cell-fate and regulatory transitions. Importantly, MAPKK and MAPK are each activated by phosphorylation on not one but two amino acid residues, full activation being attained when both residues in question

I. Maly, *Quantitative Elements of General Biology*,
https://doi.org/10.1007/978-3-030-79146-9_4

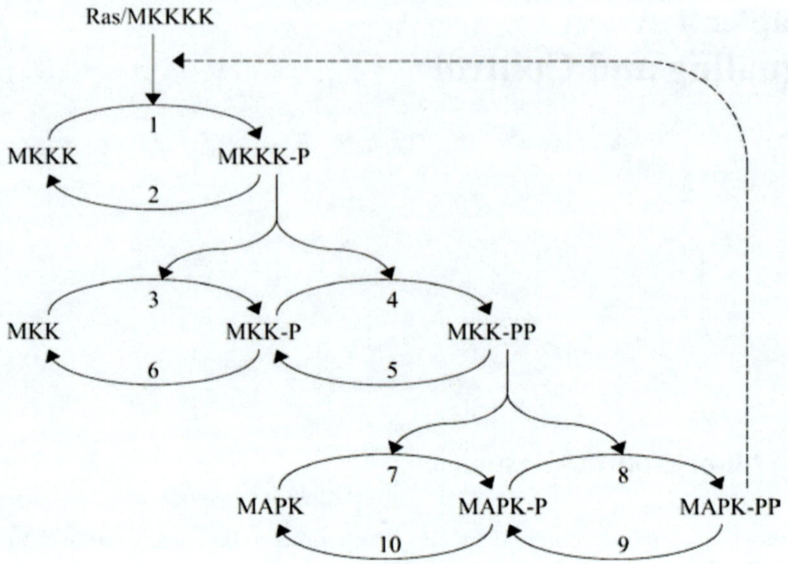

Fig. 4.1 MAPK phosphorylation cascade. Reactions are numbered. *P*, phosphorylated form, *PP*, doubly phosphorylated form. See text for other abbreviations. *Dashed line*, negative feedback considered in some model treatments. (Reproduced from Kholodenko 2000 with permission from John Wiley and Sons. Copyright FEBS 2000)

are phosphorylated on a given kinase molecule. Whether single or double, the state of phosphorylation of MAPK and MAPKK is itself dynamic, being subject to continuous reversal under the action of phosphatases (Fig. 4.1).

In experiments, when the input signal x was varied by manipulating the total concentration of MAPKKK, the levels y of activated MAPK and MAPKK were found to be well approximated by the Hill equation,

$$y = \frac{x^n}{K + x^n},$$

with the Hill coefficient n equal to 4.9 for MAPK and 1.7 for MAPKK (Fig. 4.2). The Hill equation is normally applicable to cooperative enzymes, in which several substrate-binding sites interact through the enzyme's molecular structure. Hill coefficients larger than 1 in that case result from mutual enhancement of binding of the substrate to the different sites on the same enzyme molecule. The originally studied and paradigmatic cooperative molecule, hemoglobin, is able to achieve a Hill coefficient of 2.8 when all four of its sites binding molecular oxygen become occupied. Neither MAPK nor MAPKK, however, are cooperative enzymes in this structural biophysical sense. Instead, their individual kinetics can be described by the Michaelis-Menten equation, which is equivalent to the Hill equation with $n = 1$ (Fig. 4.2).

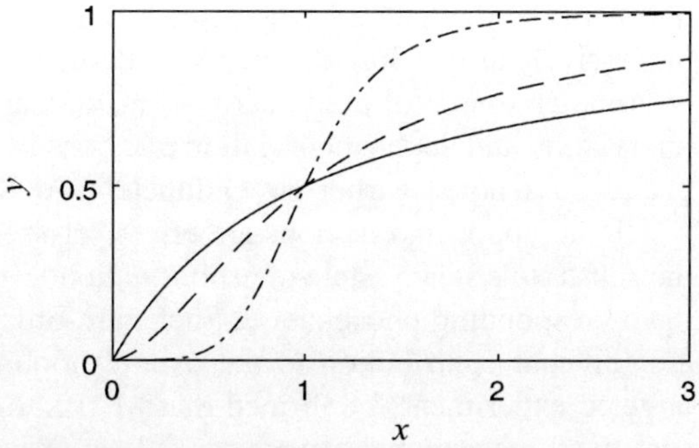

Fig. 4.2 Hill functions with $K = 1$ and $n = 1$ (*solid line*), 1.7 (*dashed line*), and 4.9 (*dash-dot line*)

In search of the origin of the apparently cooperative kinetics exhibited by the cascade, kinetics of the sequential phosphorylation and dephosphorylation reactions was analyzed in a system of binding, dissociation, product formation, and conservation equations that totaled 25. It was assumed that the readout signal from the cascade represents the steady-state fraction of the doubly phosphorylated downstream kinase, while the individual phosphorylation and dephosphorylation reactions continue after the balance has been reached. The analysis demonstrated that whether the input to the cascade was varied as the activity of an enzyme converting MAPKKK to the form capable of phosphorylating MAPKK (which would happen after the arrival of an extracellular signal) or as the total MAPKKK concentration (as was done in the in-vitro experiment), the signal readout on the levels of MAPKK and MAPK followed a nonlinear response curve that could be closely approximated by the phenomenological Hill equation. Moreover, the experimentally determined coefficient values 1.7 and 4.9, respectively, were found consistent with the measured and estimated values of the kinetic parameters in the cascade. Higher and lower Hill coefficients could be obtained depending on the exact assumptions made, but the sigmoidal shape of the response curve was a general feature of the cascade kinetics, even though no individual enzyme was assumed to be cooperative.

The main source of the additional and cumulative nonlinearity in the cascade was identified as the two-step phosphorylation at each of the downstream levels, MAPKK and MAPK. In principle, MAPKK and MAPK could be activated via a mechanism in which the activating enzyme would bind the downstream kinase and phosphorylate it at both residues in sequence before dissociating from the product. Simulations show that such a mechanism would greatly diminish the emergent cooperativity in the cascade. Indeed, in the limit of small signals it is easy to see how nonlinearity may arise in the two-step compared with the processive double phosphorylation. The singly phosphorylated product coming out of the first step would then appear at a rate proportional to the signal at the input to this level in the cascade, and serve as a mass-action reagent in the second step, whose specific rate would again be proportional to the signal from the higher level in the cascade. The combined result would be second-order kinetics of the output from the given level of the cascade with respect to the input signal. Targeted experiments showed that phosphorylation in the MAPK cascade indeed proceeds in discrete steps rather than processively after a single collision with the activating enzyme.

Another source of nonlinearity is near-saturation by any substrate of the phosphorylation and dephosphorylation reactions of the corresponding kinase and phosphatase. As noted earlier by Goldbeter and Koshland (1981), this condition, in which the opposing reactions are effectively zero-order, leads to ultrasensitivity of the substrate's steady-state phosphorylation level to the ratio of the active kinase to the corresponding phosphatase. Such zero-order ultrasensitivity was found to have a significant contribution to the overall nonlinearity of the MAPK cascade, and targeted experiments confirmed that MAPK Erk2 in the system where the signal sensitivity measurements were carried out (*Xenopus* egg extracts) was indeed present at saturating concentrations. Interestingly, while yeast MAPK and MAPKK occur at concentrations that are up to two orders of magnitude lower, the MAPK-MAPKK affinity in yeast is two orders of magnitude higher, suggesting that the zero-order ultrasensitivity may be as conserved across different clades as the molecular components of the MAPK cascade themselves. This might be expected given the cascade's function of triggering irreversible decisions about the cell fate (to divide or differentiate, for example), in which the ability to filter out weak signals and transform stronger ones into uniform outputs is critical.

While nonlinearity of the signal transformation by the MAPK cascade is considerable, on close inspection with sensitive methods it falls short of an all-or-nothing response. A class of responses mediated by ion fluxes across the cell membrane come closer to that theoretical notion and are associated with secretory functions in diverse contexts, from digestive to neurobehavioral. As shown by Hodgkin and Huxley (1952) in their work on axons, activation and inactivation of sodium and potassium channels in the plasma membrane, which are dependent on the electric potential difference across the membrane, can lead to excitatory dynamics. Although it has been elaborated both with respect to the molecular details and to the supracellular physiological context, the Hodgkin-Huxley model remains at the core of modern analyses. It considers the following dynamics (in the formulation of Hansel et al. 1993 for neurons):

$$\begin{aligned} C\frac{dV}{dt} &= I - g_{Na}m^3h\cdot(V - V_{Na}) - g_{K}n^4\cdot(V - V_{K}) - g_l\cdot(V - V_l), \\ \frac{dm}{dt} &= \frac{m_\infty(V) - m}{\tau_m(V)}, \\ \frac{dh}{dt} &= \frac{h_\infty(V) - h}{\tau_h(V)}, \\ \frac{dn}{dt} &= \frac{n_\infty(V) - n}{\tau_n(V)}. \end{aligned} \tag{4.1}$$

Here, C is the capacitance density of the membrane and V is the potential difference across it (called "membrane potential"). I is the controlling current density—a current through the membrane that is external to the model and representing, for example, the current through a receptor channel binding an extracellular signaling molecule. The other terms in the first equation represent the current densities through sodium (Na) and potassium (K) channels, and the density of a nonspecific ("leak") current, in that order. The passive currents are proportional to the difference between the membrane potential and the reversal potential which is unique to each current (V_{Na}, V_{K}, V_l). The reversal potentials are the balancing potential differences that correspond to the differences in each specific ion's concentration across the membrane, or that of the mix of ions responsible for the leak current. (The ion concentrations are maintained out of equilibrium by active transporters outside the model system and can be assumed not to be changed appreciably by the current dynamics that is the model's subject.)

The maximum conductivity associated with each of the three passive currents in Eq. (4.1) is denoted by g. In the case of the leak current, this is the constant leak conductivity g_l. The sodium and potassium conductivities, however, are modulated by the channel activation and deactivation variables, h, m, and n. The evolution of each of these variables is controlled by two functions of V, one of which represents the equilibrium value that the variable would reach at the specified membrane potential, and the other the time constant of the variable's approach to the equilibrium value. The said functions are phenomenological and resulted from Hodgkin and Huxley's analytical approximations of experimental measurements. Their complete notation is worth giving here instead of the frequently encountered shorthand. Each equilibrium value is in the form $a/(a + b)$, and each time constant is in the form $1/(a + b)$, where a and b are as follows:

$$a_m = \frac{0.1\mathrm{ms}^{-1}\mathrm{mV}^{-1}\left(40\mathrm{mV}+V\right)}{1-\exp\left(\dfrac{-40\mathrm{mV}-V}{10\mathrm{mV}}\right)},\quad b_m = 4\mathrm{ms}^{-1}\exp\left(\frac{-65\mathrm{mV}-V}{18\mathrm{mV}}\right),$$

$$a_h = 0.07\mathrm{ms}^{-1}\exp\left(\frac{-65\mathrm{mV}-V}{20\mathrm{mV}}\right),\quad b_h = \frac{1\mathrm{ms}^{-1}}{1+\exp\left(\dfrac{-35\mathrm{mV}-V}{10\mathrm{mV}}\right)},$$

$$a_n = \frac{0.01\mathrm{ms}^{-1}\mathrm{mV}^{-1}\left(55\mathrm{mV}+V\right)}{1-\exp\left(\dfrac{-55\mathrm{mV}-V}{10\mathrm{mV}}\right)},\quad b_n = 0.125\mathrm{ms}^{-1}\exp\left(\frac{-65\mathrm{mV}-V}{80\mathrm{mV}}\right).$$

In addition to the phenomenological constants found in the above expressions, the following values have been adopted for the parameters in Eq. (4.1): $C = 1\mu\mathrm{F/cm}^2$, $V_{\mathrm{Na}} = 50$ mV, $V_{\mathrm{K}} = -77$ mV, $V_l = -54.4$ mV, $g_{\mathrm{Na}} = 120$ mS/cm^2, $g_{\mathrm{K}} = 36$ mS/cm^2, $g_l = 0.3$ mS/cm^2.

The membrane potential model with the selected parameters relaxes to a transmembrane potential difference of −65 mV in the absence of the external signal ($I = 0$). Application of a small control current causes a perturbation leading to a membrane potential response of a few mV, which is damped in an oscillatory

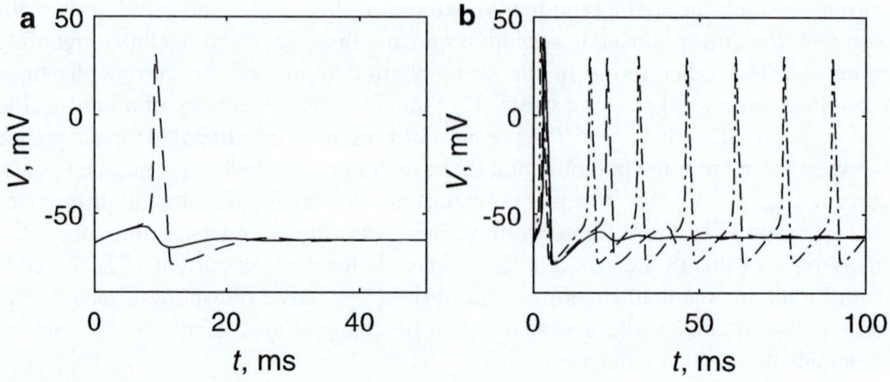

Fig. 4.3 Membrane potential in the Hodgkin-Huxley model. (**a**) *solid line*, $I = 2.24\mu A/cm^2$, *dashed line*, $I = 2.25\mu A/cm^2$, (**b**) *solid line*, $I = 5\mu A/cm^2$, *dashed line*, $I = 6\mu A/cm^2$, *dash-dot line*, $I = 10\mu A/cm^2$

fashion within tens of milliseconds (Fig. 4.3a). However, a threshold stimulus value exists around $2.245\mu A/cm^2$ above which the system exhibits a large spike in the transmembrane potential following the application of the stimulus. The depolarization spike is so large that it overshoots the zero potential and causes a brief reversal of the electric polarity on the membrane. The abrupt transition to the excitatory response (Fig. 4.3a) is an outstanding example of the all-or-nothing type of response function displayed by biological systems.

It can be additionally noted that the damped oscillations that follow the spike are transformed, when the control current exceeds further thresholds, into a pattern with an additional spike or spikes of almost the same magnitude, and just below the driving current of $10\mu A/cm^2$ the system undergoes an inverted Hopf bifurcation leading to sustained oscillations (Fig. 4.3b). The distinguishing feature of this bifurcation type, known also as "canard," is that the sustained oscillations emerge at a finite amplitude. Dynamic simulations of the membrane potential, such as the parametric series shown in Fig. 4.3, and the familiar context of the excitatory-cell physiology make this behavior natural, but it may be worth noting that the behavior of this dynamic system is qualitatively different from that around an ordinary (direct) Hopf bifurcation such as encountered in the previous chapters, whereby the oscillations emerged at an infinitesimal amplitude. Crossing immediately into the region of finite sustained oscillations is the second type of all-or-nothing response exhibited by neurons. Whereas on the cell level, the neuronal physiology is often concerned primarily with the all-or-nothing transitions from small damped responses to firing and then to sustained firing, modulation of the neurons' firing rate in the sustained regime can itself form a basis for organism-level oscillatory control mechanisms to be explored later in this chapter.

4.2 Oscillatory Control

Whereas the neuron model from the last section exhibited an all-or-nothing oscillatory response via an inverted Hopf bifurcation, the MAPK phosphorylation cascade is capable of oscillatory regulation through a normal Poincaré-Andronov-Hopf transition. This behavior was studied by Kholodenko (2000) and can serve as a simple biochemical paradigm for compartmental and organism-level oscillatory control processes. The prerequisite for oscillations in the MAPK cascade is a negative feedback influence from the activated MAPK on the rate of the reaction of activation of MAPKKK (Fig. 4.1). Biochemically, the non-competitive inhibition occurs when activated MAPK phosphorylates SOS, a protein that serves as an activator of Ras, which in turn functions by recruiting MAPKKK to the plasma membrane, where the first step of the MAPK cascade takes place. The Michaelis-Menten kinetics of the cascade phosphorylation and dephosphorylation reactions can be modeled as

$$v_1 = \frac{V_1 m_1}{\left(1 + m_3^{pp} / K_I\right)\left(K_1 + m_1\right)}, \quad v_2 = V_2 m_1^p / \left(K_2 + m_1^p\right),$$

$$v_3 = k_3 m_1^p m_2 / \left(K_3 + m_2\right), \quad v_4 = k_4 m_1^p m_2^p / \left(K_4 + m_2^p\right),$$

$$v_5 = V_5 m_2^{pp} / \left(K_5 + m_2^{pp}\right), \quad v_6 = V_6 m_2^p / \left(K_6 + m_2^p\right),$$

$$v_7 = k_7 m_2^{pp} m_3 / \left(K_7 + m_3\right), \quad v_8 = k_8 m_2^{pp} m_3^p / \left(K_8 + m_3^p\right),$$

$$v_9 = V_9 m_3^{pp} / \left(K_9 + m_3^{pp}\right), \quad v_{10} = V_{10} m_3^p / \left(K_{10} + m_3^p\right),$$

where m_1, m_2, m_3 are the three kinases forming the cascade (*i.e.*, MAPKKK, MAPKK, and MAPK), and k_n, V_n, and K_n are respectively the kinetic constant, maximum rate, and Michaelis-Menten constant of reaction n, numbered as in Fig. 4.1. The superscripts p and pp denote the phosphorylated and doubly phosphorylated forms of each kinase, while K_I is the inhibition constant. The dynamics of the kinase concentrations is then given by

$$dm_1 / dt = v_2 - v_1, \quad m_1^p = m_1^{tot} - m_1,$$

$$dm_2 / dt = v_6 - v_3, \quad dm_2^p / dt = v_3 - v_4 + v_5 - v_6, \quad m_2^{pp} = m_2^{tot} - m_2 - m_2^p,$$

$$dm_3 / dt = v_{10} - v_7, \quad dm_3^p / dt = v_7 - v_8 + v_9 - v_{10}, \quad m_3^{pp} = m_3^{tot} - m_3 - m_3^p,$$

where m^{tot} are the total conserved concentrations of the kinases.

Numerical analysis shows that given a sufficient constant input to the cascade (in the form of V_1), the cascade output, m_3^{pp}, displays steady-state oscillations in a wide range of parameter values (Fig. 4.4). It is also possible to demonstrate analytically that the bifurcation leading to the emergence of the cycles occurs in a generalized cascade of at least three reversible reactions when the strength of the negative feedback exceeds a certain value (Kholodenko 2000).

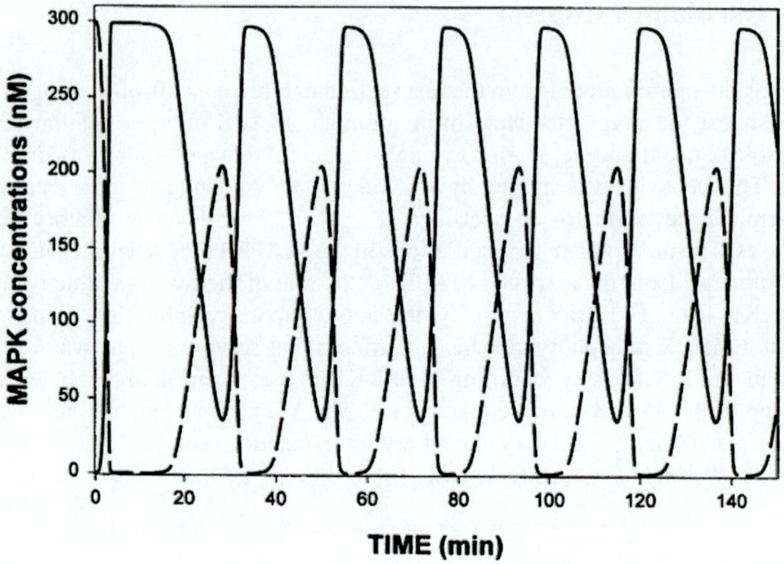

Fig. 4.4 Oscillations in the MAPK phosphorylation cascade. *Solid line*, doubly phosphorylated MAPK, *dashed line*, dephosphorylated MAPK. (Reproduced from Kholodenko 2000 with permission with permission from John Wiley and Sons. Copyright FEBS 2000)

Biochemical oscillations were documented early in glycolysis, and, as was reviewed already by Hess and Boiteux (1971), a number of diverse biochemical systems exhibit metabolism- and cell division-related oscillations, many of which include ion fluxes across plasma and organelle membranes (also outside the nervous system) and gene expression modulation. The theoretical role of feedback mechanisms in systems of this type was similarly highlighted early (Dibrov et al. 1982). Complex oscillatory systems, such as the one controlling the biochemistry of cell division, can involve both negative and positive feedback loops (Novak and Tyson 1993). At the time of its creation, the MAPK cascade oscillation model expounded here constituted a prediction, but both the oscillations (including oscillatory regulation of gene expression downstream of the cascade) and the critical role of the negative feedback from MAPK to SOS have since been experimentally confirmed in studies of signaling induced by engagement of various growth factor receptors (Nakayama et al. 2008; Shin et al. 2009). More detailed theoretical studies of oscillations in this signal transduction network continue (Arkun and Yasemi 2018).

On the organism level, oscillatory control is exemplified by the recently elaborated dynamic systems properties of the brain stem control of breathing. Although the most detailed quantitative investigations have focused on the mammalian system, gill breathing control in taxa as distant as Cyclostomata (Russell 1986) is similar if perhaps only partially homologous. The mammalian models are presently capable not only of capturing the central features of the neuroanatomy, neurophysiology, and physiology of the peripheral breathing apparatus, but also of accounting for the effects of specific experimental interventions and correctly predicting new dynamic properties (Molkov et al. 2017). One approach in these models has been based on operating with populations of Hodgkin-Huxley neurons, each modeled on the molecular level along the lines described in the last section and then connected to capture the known architecture of the breathing centers in the brain stem. The system-level results of this approach, however, can be captured successfully and further analyzed in non-spiking models that operate with unitary elements that each account for the activity of an entire brain stem nucleus or a physiologically defined neuronal population within the given breathing control region of the brain stem. This second approach has been termed "activity-based" and will be the one expounded in the present section.

Each neuron population i in the activity-based model of breathing control (Molkov et al. 2014) is represented by its transmembrane electric potential difference V_i that evolves according to a modified Hodgkin-Huxley equation:

$$\begin{aligned} C\frac{dV_i}{dt} &= -g_K n^4 \cdot (V_i - V_K) - g_l \cdot (V_i - V_{li}) \\ &-I_i - g_E q_i \cdot (V_i - V_E) - g_I s_i \cdot (V_i - V_I). \end{aligned} \tag{4.2}$$

Here, C, as before, captures the effect of capacitance. However, compared to the spiking Hodgkin-Huxley model considered earlier (Eq. 4.1), the breathing control model is formulated in terms of cell populations rather than the plasma membrane areas, and so C now has the dimensions of capacitance instead of capacitance density. The first two terms on the right-hand side, which represent the potassium rectifier current and leak current, retain the forms they had in Eq. (4.1), the g factors now having the dimensions of conductance. In comparison, however, the treatment of the potassium current is simplified by an assumption that inactivation n follows the transmembrane potential instantaneously:

$$n(V_i) = \frac{1}{1 + \exp\left(\frac{-V_i - 30\text{mV}}{4\text{mV}}\right)}.$$

The last three terms in Eq. (4.2) represent, correspondingly, the neuron population-specific current and the excitatory and inhibitory synaptic influences. The form of these is best understood with reference to the functional connections among the relevant brain stem neuron populations. The cycle of neural activity underlying the breathing rhythm (Fig. 4.5) arises from the spontaneously cycling activity of pre-inspiratory/inspiratory (referred to as "pre-I/I") neurons, and is orchestrated by inhibitory influences from and among the early inspiratory (early-I), post-inspiratory (post-I), and augmenting expiratory (aug-E) neurons, which also process tonic inputs from the pons and the retrotrapezoid nucleus (RTN). Late expiratory (late-E) neurons coordinate in a bidirectional excitatory and inhibitory manner with several of these core network components, and are capable of adding a second cyclical drive under non-basal physiological conditions. The normal inspiratory signal is smoothly

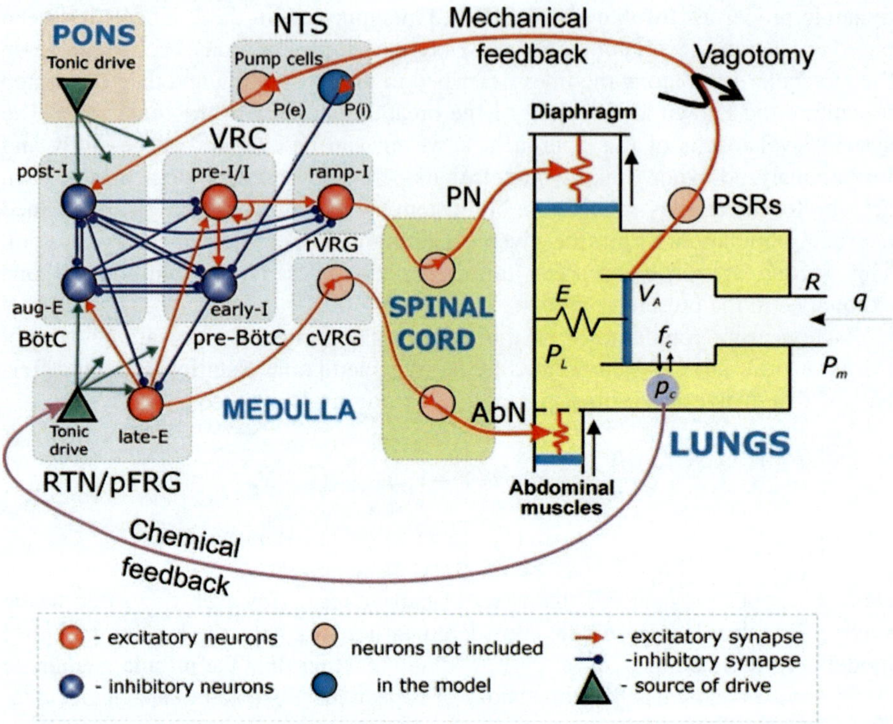

Fig. 4.5 Brain stem control of breathing. *AbN*, abdominal nerve, *BötC*, Bötzinger complex, *cVRG*, caudal ventral respiratory group, *NTS*, nucleus of the tractus solitarius, *P(e)*, excitatory pump cells, *P(i)*, inhibitory pump cells (note that these are not explicitly modeled but relay the signal denoted *L* in the equations), *pFRG*, parafacial respiratory group, *PN*, phrenic nerve, *PSR*, pulmonary stretch receptor, *rVRG*, rostral ventral respiratory group, *VRC*, ventral respiratory column. Other abbreviations are explained in the text. (Reproduced from Molkov et al. 2014 with permission)

"ramped up" by neurons denoted as ramp-I (Fig. 4.5), which integrate the stimulatory and inhibitory signals from the four core neuron populations.

The spontaneous cycling drive in the pre-I/I and late-E neurons ($i = 1$ and 2, correspondingly) arises in connection with the persistent sodium (NaP) current, which is modeled as follows:

$$I_i = g_{\mathrm{NaP}} m(V_i) h_i \cdot (V_i - V_{Na}) \quad (i = 1,2),$$

$$m(V_i) = \frac{1}{1 + \exp\left(\dfrac{-V_i - 40\mathrm{mV}}{6\mathrm{mV}}\right)}, \quad \tau_h(V_i)\frac{dh_i}{dt} = h_\infty(V_i) - h_i,$$

$$\tau_h(V_i) = \frac{4\mathrm{ms}}{\cosh\left(\dfrac{V_i + 55\mathrm{mV}}{10\mathrm{mV}}\right)}, \quad h_\infty(V_i) = \frac{1}{1 + \exp\left(\dfrac{V_i + 55\mathrm{mV}}{10\mathrm{mV}}\right)}.$$

As we can see, as far as the activation function *m*, the breathing model follows the time-separation approach already employed in the approximation adopted for the potassium current, compared with the full Hodgkin-Huxley model. The inactivation variable *h*, however, is modeled in a time-dependent fashion.

For the early-I, aug-E, and post-I neurons ($i = 3, 4, 5$), I_i represents the adaptive potassium current:

$$I_i = g_{KA} a_i \cdot (V_i - V_K),$$
$$\tau_{ai} \frac{da_i}{dt} = K_i f(V_i) - a_i \qquad (i = 3,4,5).$$

Here, *f* is the neuron activity function:

$$f(V_i) = \begin{cases} 0, & \text{if} V_i < -50\text{mV} \\ (V + 50\text{mV}) / 30\text{mV}, & \text{if} -50\text{mV} \le V_i \le -20\text{mV} \\ 1, & \text{if} V_i > -20\text{mV} \end{cases}$$

One can see that the activity function effectively stands for the time-averaged transmembrane potential that would be tracked as a measure of the neuronal population's activation in a more detailed, spiking Hodgkin-Huxley model. Finally, defining the specific current I_i as zero for the ramp-I neuron population ($i = 6$) completes the description as far as this term in Eq. (4.2) is concerned.

The synaptic connectivity terms in Eq. (4.2) involve the excitatory (*q*) and inhibitory (*s*) gating variables derived as follows:

$$q_i = \sum_{j=1}^{5} \alpha_{ij} f(V_j) + \sum_{j=1}^{2} \beta_{ij} D_j + \delta_i L,$$
$$s_i = \sum_{j=1}^{5} \gamma_{ij} f(V_j) + \varepsilon_i L.$$

The Greek notation here is for the synaptic weights, which are set to zero if the corresponding connection is not considered in the model (Fig. 4.5). D_j is the driving stimulus from the pons ($j = 1$) or RTN ($j = 2$), and *L* is the variable that tracks the feedback strength from lung stretch receptors. D_1 is taken to be equal to 1, while D_2 is modeled as a hyperbolic tangent function of the partial CO_2 pressure in the blood, as perceived by chemoreceptors. *L* is proportional to the inspired volume, *i.e.*, the lung volume above the basal one.

The details of the gas exchange and transport, as well as the breathing mechanics lie outside our focus on control mechanisms, except that it should be mentioned that the model can account both for the inspiratory diaphragm contraction followed by passive exhalation during normal breathing, and for the active exhalation under the action of abdominal muscles that occurs when the metabolic load is high (Fig. 4.5). The reader may be referred to the cited original literature for all the empirically determined parameter values in the model, but it is worth noting that although all the synaptic weights in the model are positive, the excitatory and inhibitory synaptic terms in Eq. (4.2) are of the opposite sign in the model's intended regime. This is due to the fact that the corresponding reversal potentials, V_E and V_I, bracket the range of transmembrane potentials to which the model (specifically, activity function f) is sensitive.

Oscillations emerge in the model as a result of the interactions of the central neuron populations, and lead to establishment of a stable breathing rhythm (Fig. 4.6).

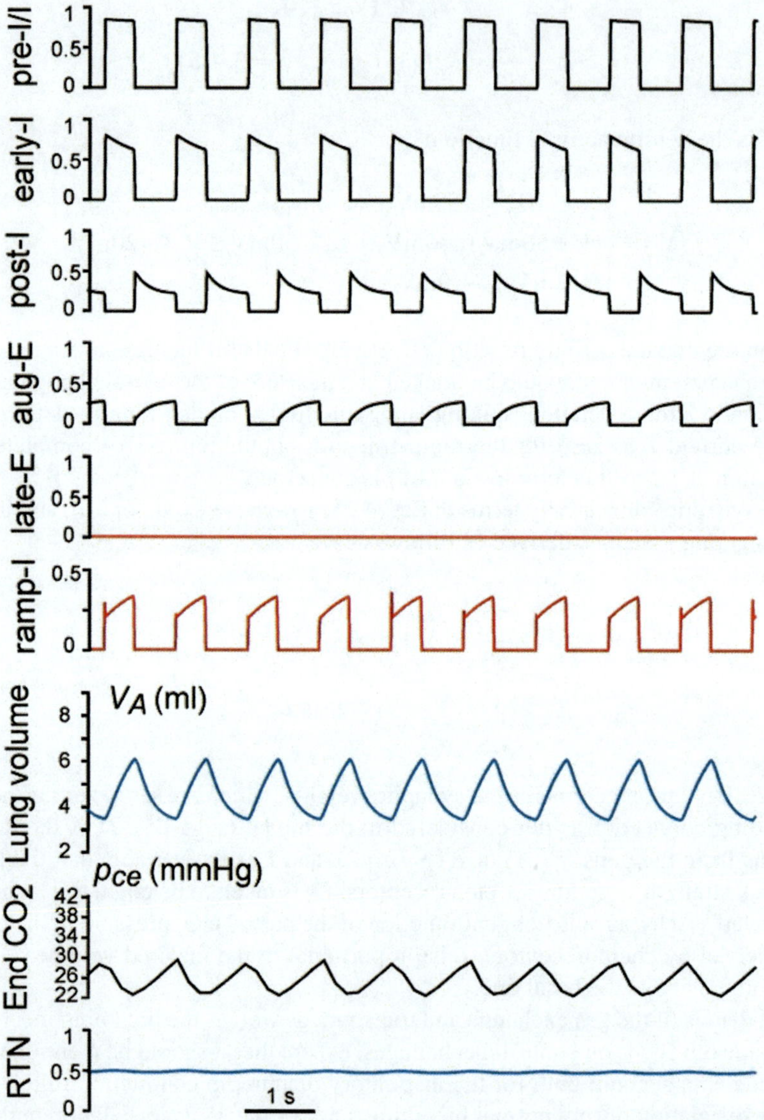

Fig. 4.6 Breathing model output. *End* CO_2 is a specific measure of the blood partial CO_2 pressure. The units refer to rat. (Reproduced from Molkov et al. 2014 with permission)

Activity of the pre-I/I neurons (the first trace in the figure) rises when the inhibition from the last cycle wanes. This activates the ramp-I and early-I neurons. (See the connectivity diagram in Fig. 4.5 and subsequent traces in Fig. 4.6.) The early-I neurons are inhibitory and act on the post-I and aug-E neurons, causing diminution of the two latter populations' inhibitory influence on ramp-I. The result of the direct and the relayed double-negative activation is the steady rise of the ramp-I output, which drives the smooth inhalation via the diaphragm contraction. The lung volume rises and the partial pressure of CO_2 in the capillaries falls (traces 2 and 3 from the bottom in Fig. 4.6). Adaptation in the early-I neurons diminishes these neurons' inhibitory effect and eventually permits activation of the post-I neurons. The latter neuron population suppresses all inspiratory neurons, setting off the phase of expiration. Adaptation in early-I and rising inhibition of this population from aug-E together eventually suppress early-I sufficiently for the release of its inhibitory influence on pre-I/I, completing the cycle.

Note that throughout the described cycle of quiet breathing, the RTN drive, which is coupled to the nonlinear output from the CO_2 chemoreceptors, remains at a moderate level (last trace in Fig. 4.6). This characterizes the conditions of low metabolic demand in the organism and determines the silence of the late-E neurons (middle trace in same figure). Correspondingly, the output through the abdominal nerve (Fig. 4.5) is absent, and breathing is actuated entirely through the diaphragm and tissue elasticity. Challenging the model to emulate the conditions of insufficient ventilation (hypercapnia), which arise under elevated metabolic load but can also be induced by elevating the outside-air CO_2 concentration, leads to a rise in the blood partial CO_2 pressure (Fig. 4.7). The RTN drive becomes correspondingly elevated, its influence on the core neuron populations causing a rise in both the frequency and amplitude of diaphragm-mediated breathing. At sufficient levels of hypercapnia, such as the case illustrated in Fig. 4.7, the RTN input to the late-E neurons additionally causes their activity to spike at the end of every n-th breathing cycle. The late-E signal is relayed to the abdominal muscles, causing an additional volume to be exhaled following the passive exhalation phase in the cycle (middle traces in Fig. 4.7). The cycle ratio n depends inversely on the level of hypercapnia, and becomes 1 after a certain threshold. The late-E spikes are locked in phase to the overall cycle due to the cyclic influences on late-E from the core inhibitory neurons (refer to Fig. 4.5). Overall, it can be observed that the breathing control accounting for the chemoreception of CO_2 under strenuous conditions takes on a multiperiodic (variable diperiodic) character, which degenerates back to uniperiodic when hypercapnia becomes severe.

We have encountered multiperiodicity already in biomass kinetics, and will again see it in widely different biological systems in the subsequent chapters. On the intracellular level, one of the best-studied control systems exhibiting multiperiodic dynamics is the calcium system. Cells maintain a low cytosolic calcium concentration via active mechanisms expelling calcium ions both across the plasma membrane and into the lumen of organelles such as endoplasmic reticulum. In addition, calcium can be sequestered in mitochondria due to these organelles high transmembrane electric potential difference. This system has functional readouts such as

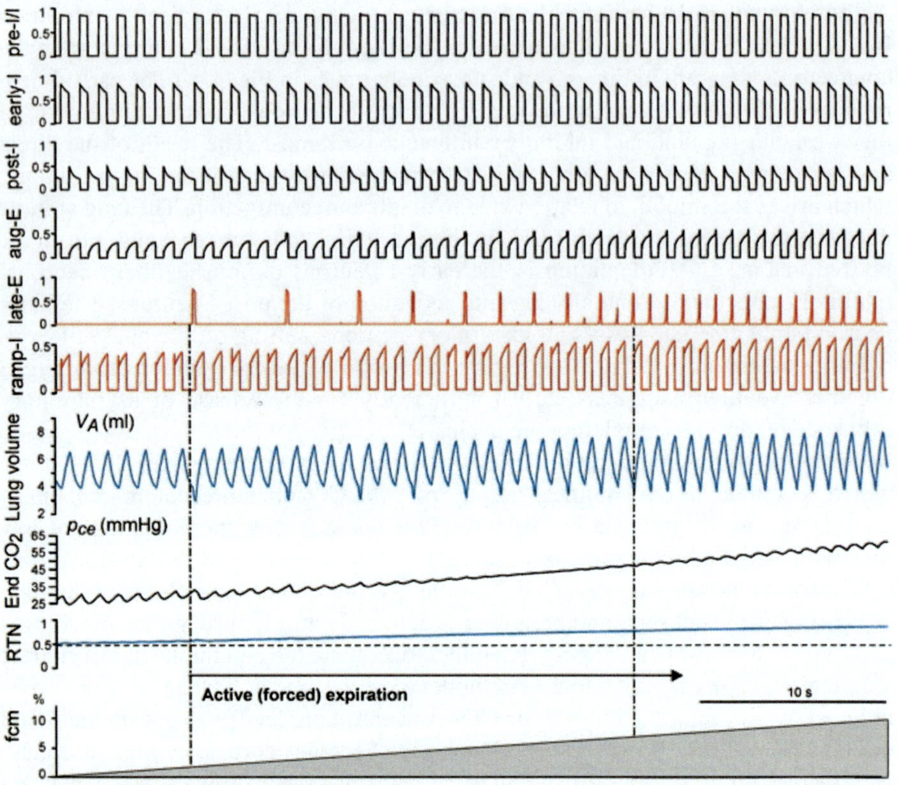

Fig. 4.7 Breathing model output under hypercapnia. The bottom trace shows the outside-air CO_2 concentration. (Reproduced from Molkov et al. 2014 with permission)

calcium-regulated import of transcription factors into the nucleus (see Maly and Hofmann 2016; Maly and Hofmann 2018; Maly and Hofmann 2020), and its dynamics is modulated by extracellular signals transmitted through plasma membrane receptors. Participation of intracellular (primarily cytosolic) calcium in transmission of the extracellular signals is why this ion has been termed a "second messenger." However, it is more accurate to characterize the extracellular signals as modulators of intrinsic dynamics of the intracellular calcium system. Kinetic analysis of the calcium system and its emergent properties remains a very active area of research (e.g., Fall et al. 2002; Cheng and Zaikin 2020). Self-organization of rich dynamic behavior such as the multiperiodicity and apparent aperiodicity in the calcium signaling system is of particular interest from the general biological viewpoint.

An illuminating analysis of multiperiodic and aperiodic oscillations in the intracellular calcium system has been offered by Haberichter et al. (2001). The total cellular concentration of free calcium ions is considered as comprised of the cytosolic concentrations of free and protein-bound calcium ions, as well as the calcium concentration in the endoplasmic reticulum and mitochondria. This is expressed by the corresponding terms in the following conservation relationship:

$$c_t = c_c + c_p + \frac{\rho_e}{\beta_e} c_e + \frac{\rho_m}{\beta_m} c_m.$$

The numerators, ρ, denote the volume ratios of the respective organelles and cytosol, while the denominators, β, denote the ratios of free to protein-bound calcium inside each type of organelle. Thus, while the model aims to capture the kinetics of calcium binding to proteins in the cytosol explicitly, binding inside the organelles is assumed to be more rapid and far from saturation. A second conservation equation relates the occupied and free calcium-binding sites on proteins in the cytosol to the total cytosolic concentration of such sites:

$$c_p + p = p_t.$$

The mitochondrial calcium concentration is controlled in this idealization by only two fluxes, J_{in} and J_{out}:

$$\frac{dc_m}{dt} = \frac{\beta_m}{\rho_m}\left(J_{in} - J_{out}\right).$$

However, these fluxes have comparatively complicated kinetics, reflective of their strong nonlinearity with respect to cytosolic calcium, which is known from experiments:

$$J_{in} = k_{in} \frac{c_c^8}{K_{in}^8 + c_c^8},$$

$$J_{out} = \left(k_{out} \frac{c_c^2}{K_{out}^2 + c_c^2} + k_m \right) c_m.$$

In fact, the high-exponent power law merely approximates an essentially step-like kinetics of the mitochondrial calcium uniporter that facilitates entry of calcium into the mitochondrial matrix during elevations of this ion's concentration in the cytosol.

The endoplasmic concentration, by comparison, is controlled by three fluxes, two of which are linear:

$$\frac{dc_e}{dt} = \frac{\beta_e}{\rho_e}\left(J_{pump} - J_{ch} - J_{leak}\right),$$

$$J_{pump} = k_{pump} c_c, \quad J_{ch} = k_{ch} \frac{c_c^2}{K_{ch}^2 + c_c^2}\left(c_e - c_c\right),$$

$$J_{leak} = k_{leak}\left(c_e - c_c\right).$$

Here, J_{pump} accounts for active transport of calcium from the cytosol into the lumen of the endoplasmic reticulum by the sarco-endoplasmic reticulum calcium ATPase, while J_{leak} describes passive exit of calcium back into the cytosol, down the prevailing concentration gradient ($c_e > c_c$). J_{ch}, the channel-mediated current out of the endoplasmic reticulum, follows a kinetic law that is nonlinear with respect to the cytosolic calcium, reflecting the mechanism known as calcium-induced calcium release. Due to an autocatalytic release of calcium stored in the endoplasmic reticulum, which this kinetic law implies, elevations of cytosolic calcium can be self-sustaining beyond a certain threshold amplitude, measured by K_{ch}. This is a property of both ryanodine-sensitive calcium channels and inositol trisphosphate receptor channels in the endoplasmic membrane. The latter type of channels is additionally activated by inositol trisphosphate downstream of a variety of plasma membrane receptors, whose activation can be therefore be modeled as an increase in the channel rate constant, k_{ch}. This dual—autocatalytic and external—regulation makes kinetics of the channel-mediated flux out of the endoplasmic reticulum (J_{ch}) central to temporal self-organization of intracellular calcium dynamics and its modulation by other signaling pathways.

The calcium system model is closed by an equation for the evolution of free calcium concentration in the cytosol:

$$\frac{dc_c}{dt} = J_{ch} + J_{leak} - J_{pump} - J_{in} + J_{out} - k_b c_c p + k_u c_p.$$

The last two terms represent binding and unbinding of calcium at buffering sites on cytosolic proteins.

Kinetics of the calcium system in this formulation has been analyzed for relevant values of the kinetic parameters, particularly k_{ch}, which is controlled by the strength of signaling input from plasma membrane receptors (Haberichter et al. 2001). Increase of k_{ch} over a critical value of 473 s^{-1} leads to a bifurcation resulting in loss of stability of the equilibrium calcium distribution between the compartments and emergence of a stable limit cycle. Simple oscillations characterized by a sequence of equal maxima and minima are replaced at around $k_{ch} = 1800$ s^{-1} by bursting, where a series of smaller-amplitude peaks follow each major maximum of c_c (Fig. 4.8a). Along with simple waveforms, bursts are common in experimental recordings of intracellular calcium from various cell types. Kinetics behind bursting can be understood if one considers that c_m falls at a nearly constant rate following each major peak, while c_c and c_e rise and fall reciprocally to generate the minor oscillations during the same period. This allowed Haberichter et al. (2001) to present an approximation to the complete system, in which c_m is treated not as a time-dependent variable but as a control parameter. On the corresponding bifurcation diagram (Fig. 4.8b), we observe a transition from a stable focus (see Appendix) when c_m is above approximately 0.8μM to a limit cycle below that value. Superimposing the complete system's limit cycle in the same coordinates (c_c, c_m) reveals that, indeed, to the right of approximately $c_m = 0.8$μM we have damped oscillations in c_c, whereas to the left of the said value the complete system exhibits

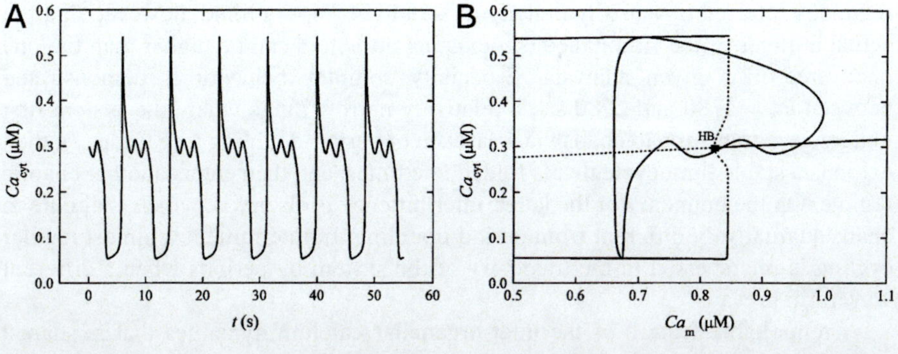

Fig. 4.8 Simple calcium bursting. (**a**) time curve of cytosolic calcium concentration, (**b**) bifurcation diagram of a reduced system in which mitochondrial calcium concentration is a control parameter, superimposed on a limit cycle of the complete system. *HB*, Hopf bifurcation. (Reproduced from Haberichter et al. 2001. Reprinted from Biophysical Chemistry, Vol. 90, T. Haberichter, M. Marhl, R. Heinrich, Birhythmicity, Trirhythmicity and Chaos in Bursting Calcium Oscillations, 17–30, copyright 2001, with permission from Elsevier)

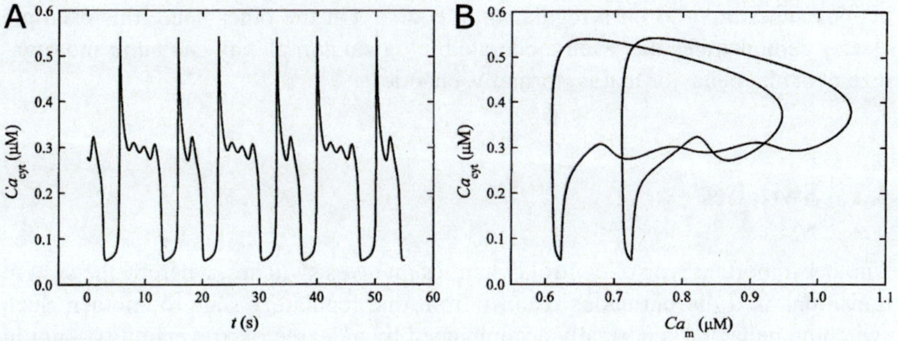

Fig. 4.9 Complex calcium bursting. (**a**) time curve of cytosolic calcium concentration, (**b**) limit cycle in the coordinates of cytosolic vs. mitochondrial calcium concentration. (Reproduced from Haberichter et al. 2001. Reprinted from Biophysical Chemistry, Vol. 90, T. Haberichter, M. Marhl, R. Heinrich, Birhythmicity, Trirhythmicity and Chaos in Bursting Calcium Oscillations, 17–30, copyright 2001, with permission from Elsevier)

oscillations of increasing amplitude. The amplitude reaches the minimum and maximum of the reduced system at points that correspond to the major minimum and maximum of the bursts (Fig. 4.8b).

With a further increase in k_{ch}, beyond 1968 s^{-1}, the system enters a birhythmic regime. In addition to the simple bursting cycle, a new stable attractor appears, which is characterized by a limit cycle that is folded, being comprised of a shorter and a longer burst (Fig. 4.9). Whether the system settles on the simple cycle or the folded one, depends on the initial conditions. At higher k_{ch}, specifically between 3225 and 3480 s^{-1}, a third attractor is found, which is also folded. The capacity of the system to settle on one of three different limit-cycle attractors in this parametric regime is referred to as trirhythmicity. It should be kept in mind, however, that the actual number of the simultaneously existing attractors can be higher than the one uncovered in a given analysis. Especially complex behavior is demonstrated between k_{ch} = 2780 and 2980 s^{-1}, a relatively narrow range where the system first undergoes a transition to chaos via a cascade of period-doubling (see Chap. 2), then exhibits a stable limit cycle that is folded five times, and then enters another chaotic regime. On the boundary of the latter, intermittency is observed, which is a route to chaos qualitatively different from period-doubling. In intermittency, almost regular cycling is interspersed in the trajectory of the system by periods when a different waveform prevails.

A remarkable feature of the inter-organellar calcium dynamics that is shaped strongly by calcium absorption into mitochondria is that the amplitudes of the major peaks in the cytosol are almost constant as their frequency and form varies. In the model considered here, the amplitude is essentially the same through the entire range of parametric regimes—even in chaos. Moreover, a pronounced characteristic frequency of fluctuations can be observed in the chaotic regimes. The example of intracellular calcium shows, on the one hand, how comparatively simple biological systems can display wholly deterministic dynamics that appears as random perturbations superimposed on a regular time course. On the other hand, this example clearly demonstrates that even moderate biological complexity can tame and organize dynamic behavior that is, formally, chaotic.

4.3 Switches

Another important type of control dynamics involves switches, whereby the system transitions in a discontinuous manner from one regulatory state to another. Such switching behavior is normally accompanied by a degree of irreversibility—not in the sense that the last state cannot be restored, but in the sense of hysteresis around the transition point. Discovered early in the history of molecular biology, the *lac* operon in *Escherichia coli* remains a model example for biological switches, and indeed for much of the homeostatic and developmental regulation of gene expression across taxa. This operon is a system of co-regulated genes involved in lactose metabolism (Jacob et al. 1960). Due to its paradigmatic status, as well as relevance to the practical needs of biotechnology, it has remained a focus of experimental and theoretical investigation (Santillán et al. 2007, ; Santillán and Mackey 2008).

The *lac* operon itself is a genomic sequence encoding β-galactosidase (the product of *lacZ*), lactose permease (the product of *lacY*), and a less metabolically important product of a third gene, *lacA* (Fig. 4.10). The expression of these genes is switched on when the bacteria have exhausted the extracellular supply of their preferred growth substrate, glucose, and allows them to transition to metabolizing a more complex sugar, lactose. One pathway leading to the *lac* operon activation is de-repression of production of cyclic adenosine monophosphate (cAMP) in the absence of a sufficient supply of glucose. Through the formation of a complex with

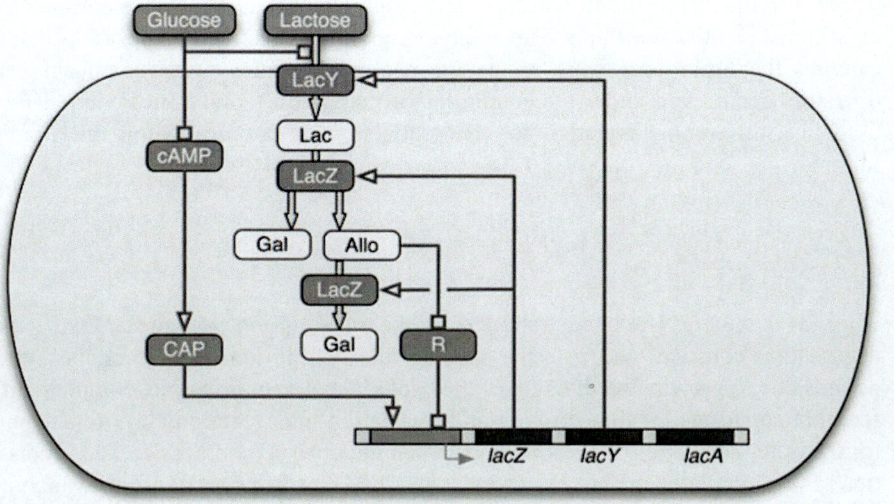

Fig. 4.10 Regulation of the *lac* operon. *Arrows*, positive influences, *squares*, negative influences. *Lac*, lactose, *Gal*, galactose, *Allo*, allolactose, *R*, repressor. See text for other abbreviations. (Reproduced from Santillán et al. 2007. Reprinted from Biophysical Journal, Vol. 92, M. Santillán, M.C. Mackey, E.S. Zeron, Origin of Bistability in the lac Operon, 3830–3842, copyright 2007, with permission from Elsevier)

catabolite activator protein (CAP), which binds the promoter region of the operon, cAMP activates binding of RNA polymerase to the operon. However, the transcription initiation cannot proceed unless repressor protein is removed from the immediately following DNA sequence called operator. The inhibition of the repressor binding requires complex formation between this protein and allolactose, which is an intermediate catabolite in the lactose pathway—specifically, a product of β-galactosidase (Fig. 4.10). In other words, a critical element of the *lac* operon activation is a positive feedback loop, which is dormant when glucose is abundant, and which becomes active when the glucose supply is low and lactose is imported into the cell by the permease and catabolized by the galactosidase. Importantly, these reactions can proceed at a sufficient rate even at the low basal (uninduced) levels of transcription in the operon to endow the cell with sensitivity to the presence of extracellular lactose.

A condensed approach to the analysis of the fundamental kinetics in the *lac* operon regulatory system has been laid out by Yildirim et al. (2004). The description accepts an empirical Hill function with the cooperativity coefficient $n = 2$ for the fraction of operator sites in the bacterial population that are unoccupied by the repressor protein:

$$F(A) = \frac{1 + K_1 A^n}{K + K_1 A^n},$$

where A is the intracellular concentration of allolactose and K, K_1 are empirically determined positive constants. The notable property of this functional form is that it captures the observed feature of the *lac* operon dynamics whereby unoccupied operators occur even under the conditions of maximum repression. Function F is used in a differential equation for the production of the *lac* operon mRNA, as follows:

$$\frac{dM}{dt} = \alpha_M F(A) - \tilde{\gamma}_M M, \qquad \tilde{\gamma}_M \equiv \gamma_M + \mu,$$

where M is the mRNA concentration, α_M is the transcription rate constant, γ_M is the degradation constant, and μ is the specific rate of dilution of intracellular compounds due to growth and division of the bacterial cells. In principle, complete differential equations for transcription and translation may feature delays to account for the time that passes, for example, between the operon derepression and production of the completed mRNA. Yildirim et al. (2004) made a special effort to analyze the delay equations and found that physiologically grounded delay times have only minor quantitative effects on the numerical results.

Proceeding with the treatment that omits the time delays, the dynamic equation for β-galactosidase (B) is analogous to the one for the operon mRNA:

$$\frac{dB}{dt} = \alpha_B M - \tilde{\gamma}_B B, \qquad \tilde{\gamma}_B \equiv \gamma_B + \mu.$$

For allolactose, we have

$$\frac{dA}{dt} = \alpha_A B \frac{L}{K_L + L} - \beta_A B \frac{A}{K_A + A} - \tilde{\gamma}_A A,$$

where L is the intracellular lactose concentration and the first two terms represent the enzymatic kinetics of allolactose production and hydrolysis, both of which are mediated by β-galactosidase. An interesting finding by Yildirim et al. (2004) is that it is possible to consider L a parameter, based on an assumption of a quasi-steady-state balance between the import and catabolism of lactose, and still account for the main dynamic features of a complete model featuring explicit kinetics of permease translation and intracellular lactose.

Analysis using parameter values estimated from experiments shows that depending on L, there may be from one to three steady-state values of A (Fig. 4.11). The intermediate value, when it exists, is unstable, and the system jumps between the low-A and high-A branches at the points where there are only two steady-state solutions. The steady state with low A signifies that β-galactosidase, and therefore the entire operon, is uninduced, and the steady state with high A corresponds to the conditions where the operon expression has been induced. The existence of the interval of L where the stable branches of the steady-state solution overlap

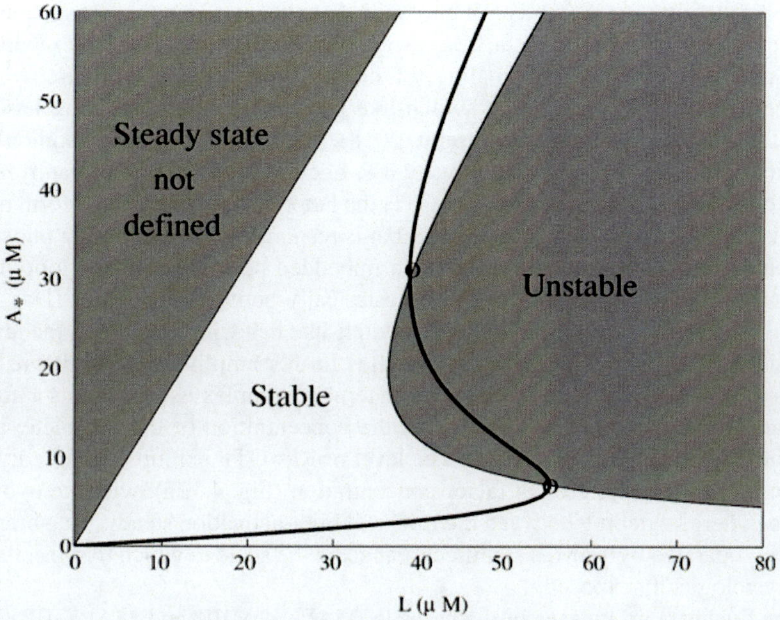

Fig. 4.11 Steady states of the *lac* operon. *Solid curve*, steady states, *shaded areas*, stability domains, *circles*, jump-off points of the switch. The model with delays is shown; when the delays are neglected, the induction point (*right circle*) is shifted to $L = 52.1\mu M$. Reprinted from Yildirim et al. (2004), with the permission of AIP Publishing

determines the presence of hysteresis in the system, whereby for these intermediate lactose concentrations, the state of induction in the *lac* operon depends on the immediately preceding state of the system: transcription that was induced stays induced, and one that was uninduced remains uninduced until the critical jump-off points are reached (Fig. 4.11). This type of dynamics enables the system to avoid metabolically inefficient states of intermediate or unstable induction and respond to significant changes in the supply of growth substrates by re-wiring the metabolism decisively in a manner that is indeed analogous to the mechanics of the traditional electric wall switch.

Eukaryotic cell signaling similarly employs switch-like dynamics, as exemplified in the cellular decision-making that is based on perception of growth factors and mediated by the already considered, central MAPK kinase cascade. To make the dynamic treatment of the growth factor perception upstream of the kinase cascade explicit, we need to consider the molecular interactions that enable the transmission of the signal from the plasma membrane receptor to the first kinase in the cascade. These critically involve sequential assembly of a Shc-Grb2-SOS protein complex on the cytoplasmic domains of activated receptors. The fully assembled complex, in the case of the well-studied and developmentally important epidermal growth factor (EGF) receptor (EGFR) facilitates the reaction of phosphorylation of Raf, a kinase that performs the function of MAPKKK. At the same time, phosphorylation of SOS, which we have already encountered in the last section, acts by sequestering this component from assembly into the ternary complex with Shc and Grb2. In the case of EGF signaling, this occurs under the action of ERK, a MAPK.

The dynamics leading to the switch-like behavior in EGF signaling has been recently analyzed by Arkun and Yasemi (2018). Their model considers explicitly the kinetics of the step whereby the active EGFR-Shc-Grb2-SOS complex catalyzes the transformation of Ras-GDP into Ras-GTP, the latter being the activated form of this plasma membrane-associated molecule that is responsible for promoting phosphorylation of Raf. This step also features an imbedded positive feedback loop due to binding of Ras-GTP to SOS, which allosterically activates the latter (Das et al. 2009). This causes a rather pronounced switch-like behavior already on the level of Raf in the EGFR signaling pathway, which is further amplified on the downstream levels (Fig. 4.12). Due to the feedback, the ternary complex itself exhibits a trace of the same switch-like behavior, as far as the concentration of the complete, active complex is concerned (Fig. 4.12a). The level of Ras-GTP exhibits a markedly hysteretic response to the growth factor concentration (Fig. 4.12b), with the two limit points (where switching between the low- and high-activation steady-state branches occurs) separated by 1 nM, a significant range on the scale to which tissue cells may be physiologically exposed.

The downstream kinases in the cascade, MAPKK MEK and MAPK ERK, display three stable branches of the steady-state concentration of their doubly phosphorylated forms (Fig. 4.12c, d). Each kinase remains essentially fully activated in a wide range of the growth factor concentration. The latter needs to drops below the left-most limit point to cause the system to jump to the state where the doubly

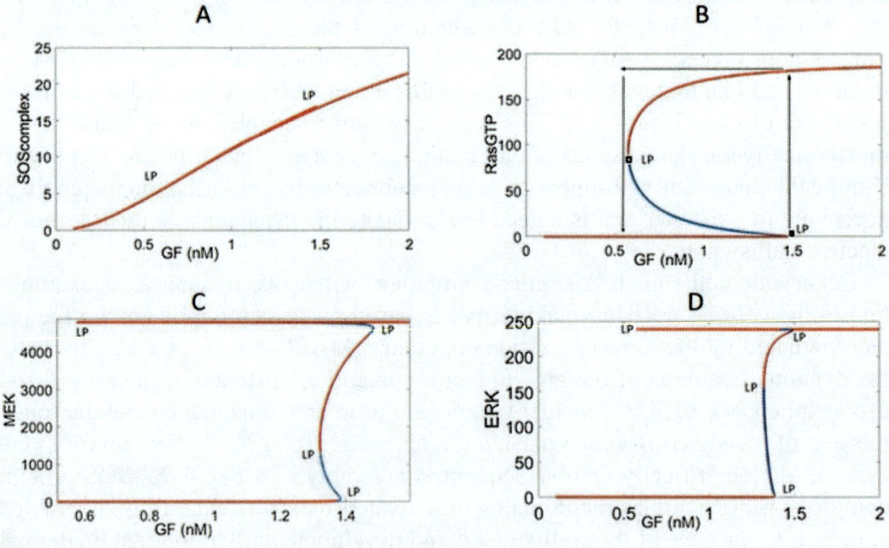

Fig. 4.12 Steady states in EGF signaling. *Red curves*, stable steady states, *blue curves*, unstable steady states, *LP*, limit point. Reproduced from Arkun and Yasemi (2018) with permission

phosphorylated form of the kinase is essentially absent. And any subsequent rise in the growth factor concentration needs to be quite significant to reach the limit point where the system abandons the inactivated branch. Unlike in the simple switching behavior that we have encountered in the previously considered systems, the MEK and ERK levels in the growth factor signaling system do not transition directly to the fully activated branch when the limit point on the inactive branch is reached (Fig. 4.12c, d). Instead, the jump leads to the intermediate stable branch, and a still higher limit point on the growth factor axis must be reached before the system transitions to the fully active branch. Although the level of activation is similar on the fully active and intermediate branches when the growth factor concentration is high, the role of the intermediate branch is revealed when the growth factor concentration decreases. If the system has not been fully activated, the decrease in the input signal drives a rapid decrease of activity on the levels of MEK and ERK, and soon causes the system to jump off to the fully inactive branch (Fig. 4.12c, d). In other words, the hysteresis is still present in the system behavior involving the inactive and intermediate stable branches, but its magnitude is insignificant compared with the one associated with the branches corresponding to the fully active and inactive states. Thus, even simplified analyses reveal complexities in the dynamic behavior of control and signal processing systems that need to be taken into account when quantifying the behavior of the model systems' real counterparts.

We have heretofore considered the controlling systems in terms of ordinary differential equations (with some involvement of time delays). The control processes in biological systems take place in physical space and so, in principle, need to be analyzed as spatially distributed systems. It is becoming clear that in many cases—whose number will undoubtedly grow as the advances in experimental methods and general maturation of quantitative biology focus the biologists' attention on spatially explicit analyses—the spatial distribution is not merely a quantitative factor playing a role in the temporal dynamics but the medium in which the primary effects arise. Although we have concentrated on the Hodgkin-Huxley dynamics of a firing neuron, the action potential model was originally derived as a model for a traveling wave propagation along an axon, and the spatial propagation of the action potential remains an area of ongoing investigation (George et al. 2015). Traveling waves that arise in populations of neurons are a promising new direction in sensory physiology and information-processing in the central nervous system (Muller et al. 2018). Phenomenologically similar waves, which have been discovered in ERK signaling on the tissue scale, accompany such processes as wound healing (Hiratsuka et al. 2015). Emergent properties of spatially distributed ERK and growth factor signaling have been explored in theoretical models on both the single-cell and complex-tissue levels, and involve spatial self-organization as well as hysteresis of the spatially differentiated activity distributions under mechanical deformation (Maly et al. 2004a, b; Tschumperlin et al. 2004). Theoretical approaches to the spatially distributed neuron populations already comprise a comparatively mature and diverse field (Muller et al. 2018). It is likely that spatially self-organized signaling will find a place among the fundamentals of perception, control, and homeostasis regulation as the experimental and theoretical research progresses.

References

Arkun Y, Yasemi M (2018) Dynamics and control of the ERK signaling pathway: sensitivity, bistability, and oscillations. PLoS One 13:e0195513

Cheng S, Zaikin A (2020) Quantitative physiology. Springer, Singapore

Das J, Ho M, Zikherman J et al (2009) Digital signaling and hysteresis characterize Ras activation in lymphoid cells. Cell 136:337–351

Dibrov BF, Zhabotinsky AM, Kholodenko BN (1982) Dynamic stability of steady states and static stabilization in unbranched metabolic pathways. J Math Biol 15:51–63

Fall CP, Marland ES, Wagner JM et al (eds) (2002) Computational cell biology. Springer, New York

George S, Foster JM, Richardson G (2015) Modelling in vivo action potential propagation along a giant axon. J Math Biol 70:237–263

Goldbeter A, Koshland DE Jr (1981) An amplified sensitivity arising from covalent modification in biological systems. Proc Natl Acad Sci U S A 78:6840–6844

Haberichter T, Marhl M, Heinrich R (2001) Birhythmicity, trirhythmicity and chaos in bursting calcium oscillations. Biophys Chem 90:17–30

Hansel D, Mato G, Meunier C (1993) Phase dynamics for weakly coupled Hodgkin-Huxley neurons. Europhys Lett 23:367–372

Hess B, Boiteux A (1971) Oscillatory phenomena in biochemistry. Annu Rev Biochem 40:237–258

Hiratsuka T, Fujita Y, Naoki H et al (2015) Intercellular propagation of extracellular signal-regulated kinase activation revealed by in vivo imaging of mouse skin. elife 4:e05178

Hodgkin AL, Huxley AF (1952) A quantitative description of membrane current and its application to conduction and excitation in nerve. J Physiol 117:500–544

Huang CYF, Ferrell JE Jr (1996) Ultrasensitivity in the mitogen-activated protein kinase cascade. Proc Natl Acad Sci U S A 93:10078–10083

Jacob F, Perrin D, Sanchez C et al (1960) Operon: a group of genes with the expression coordinated by an operator. C R Hebd Seances Acad Sci 250:1727–1729

Kholodenko BN (2000) Negative feedback and ultrasensitivity can bring about oscillations in the mitogen-activated protein kinase cascades. Eur J Biochem 267:1583–1588

Maly IV, Hofmann WA (2016) Calcium-regulated import of myosin IC into the nucleus. Cytoskeleton 73:341–350

Maly IV, Hofmann WA (2018) Calcium and nuclear signaling in prostate cancer. Int J Mol Sci 19:1237

Maly IV, Hofmann WA (2020) Myosins in the nucleus. In: Coluccio L (ed) Myosins, Advances in experimental medicine and biology, vol 1239. Springer, Cham, pp 199–231

Maly IV, Lee RT, Lauffenburger DA (2004a) A model for mechanotransduction in cardiac muscle: effects of extracellular matrix deformation on autocrine signaling. Ann Biomed Eng 32:1319–1335

Maly IV, Wiley HS, Lauffenburger DA (2004b) Self-organization of polarized cell signaling via autocrine circuits: computational model analysis. Biophys J 86:10–22

Molkov YI, Shevtsova NA, Park C et al (2014) A closed-loop model of the respiratory system: focus on hypercapnia and active expiration. PLoS One 9:e109894

Molkov YI, Rubin JE, Rybak IA et al (2017) Computational models of the neural control of breathing. Wiley Interdiscip Rev Syst Biol Med 9:e1371

Muller L, Chavane F, Reynolds J et al (2018) Cortical travelling waves: mechanisms and computational principles. Nat Rev Neurosci 19:255–268

Nakayama K, Satoh T, Igari A et al (2008) FGF induces oscillations of Hes1 expression and Ras/ERK activation. Curr Biol 18:R332–R334

Novak B, Tyson JJ (1993) Numerical analysis of a comprehensive model of M-phase control in *Xenopus* oocyte extracts and intact embryos. J Cell Sci 106:1153–1168

Russell DF (1986) Respiratory pattern generation in adult lampreys (*Lampetra fluviatilis*): interneurons and burst resetting. J Comp Physiol A 158:91–102

Santillán M, Mackey MC (2008) Quantitative approaches to the study of bistability in the *lac* operon of *Escherichia coli*. J R Soc Interface 5:S29–S39

Santillán M, Mackey MC, Zeron ES (2007) Origin of bistability in the *lac* operon. Biophys J 92:3830–3842

Shin SY, Rath O, Choo SM et al (2009) Positive- and negative-feedback regulations coordinate the dynamic behavior of the Ras-Raf-MEK-ERK signal transduction pathway. J Cell Sci 122:425–435

Tschumperlin DJ, Dai G, Maly IV et al (2004) Mechanotransduction through growth-factor shedding into the extracellular space. Nature 429:83–86

Yildirim N, Santillán M, Horike D et al (2004) Dynamics and bistability in a reduced model of the *lac* operon. Chaos 14:279–292

Chapter 5
Self-Organization in the Cell

5.1 Orientation of the Cortex

Today's biology is exploring the functional architecture of the cell as a complex structure consisting of various molecules. The molecules are not simply bound to each other but also dynamically interact, so that the supramolecular structure is observed in the state of continuous, rapid rebuilding. As our knowledge of the molecules and their own submicroscopic structure grows, it becomes an increasingly pressing problem to understand the origin of the large-scale order in biomolecular ensembles. How is the large-scale structure determined by the kinetics of interaction of its molecular components? In this chapter we will address this general problem through the prism of quantitative models of cytoskeleton dynamics, cytoskeleton-directed and -driven motility, and spatially distributed cell signaling. In the cytoskeleton dynamics and cell motility, the kinetic origin of the functioning structure is the most evident. Active processes of polymerization of cytoskeletal proteins into cell-spanning fibers of functionally significant orientation and length, which undergo continuous remodeling, and transport of organelles along these fibers that generates, maintains, and alters the organelles' distribution in the cell, together provide the dynamic structural basis for spatial organization of the entire cell. Recent years have seen rapid progress in theoretical and experimental studies across the entire spectrum of the cytoskeleton dynamics and cell motility, which has led to a certain crystallization of the main themes of emergent behavior in these spatially distributed dynamic systems, even if our understanding the most fundamental problems of the physical origin of eukaryotic cell structure on the quantitative level can still be characterized as nascent.

We have already considered, in an earlier chapter, the dynamic behavior stemming from the property that makes the higher-level biological systems capable of self-organization, namely that such systems consist of proliferating entities that inherit traits influencing their survival and reproduction. Regardless of the nature of the entities being selected, these are sufficient conditions for natural selection, as

I. Maly, *Quantitative Elements of General Biology*,
https://doi.org/10.1007/978-3-030-79146-9_5

reviewed by Michod (1999). Indeed, not only speciation, but also development of immune and neural networks (Rowe 1994) and communication (Nowak et al. 2000; Nowak et al. 2001), among other phenomena, have been studied in an evolutionary framework. Of special interest in the present context is that self-organization by means of natural selection is also one of the mechanisms that shape the architecture of an individual cell directly, in the process of that cell's individual development, and influence the aspects of the cell architecture that may not be inherited from the mother cell (Maly and Borisy 2001). This perspective is based on the recognition of polymerizing actin filaments in the dynamic cell cortex, and their growing assemblies, as entities subject to immediate natural selection on the intracellular time scale.

The cortical actin cytoskeleton is an example of a supramolecular complex whose structure regenerates at a high rate. Its dynamics plays a central role in protrusive activity of a typical eukaryotic cell's boundary, and is most obvious at the leading edge of directionally crawling cells. An elastic Brownian ratchet model has been elaborated for the cortical actin filaments (Mogilner and Oster 1996), which considers how these filaments exert force on the plasma membrane and convert the free energy expended in their non-equilibrium polymerization into expansion of the cell boundary. In a precise relationship determined by the chemical physics of the filament's abutment at the membrane, the force exerted depends on the abutment angle. Orientation of the filaments in the propulsive actin array is therefore an important factor of both cell locomotion and morphogenesis that result from the cell boundary deformation. The modal abutment angle of actin filaments in the best-studied metazoan systems has long been known to be in the range of 25 to 45°, the values to whose mechanistic significance we will return shortly.

The filaments in the cortical actin network are connected through end-to-side junctions which, in the network growth dynamics, represent the points of nucleation of a new filament on an existing one. Such dendritic pattern of polymerization nucleation is determined by the Arp2/3 protein complex that nucleates assembly of new filaments and subsequently remains at the junctions, defining the angle between the two filaments through the geometry of the ternary actin-Arp2/3-actin molecular interactions. The angle between the filaments joined by Arp2/3 is termed branching angle and comprises approximately 70° (Svitkina and Borisy 1999). Dendritic nucleation takes place in the immediate proximity of motile sections of the plasma membrane, such as the lamellipodial leading edge, where Arp2/3 is recruited by Scar and activated by WASP (Wiskott-Aldrich syndrome protein).

The assembled array of filaments is disassembled deeper in the cortex to recycle its components to the outer boundary, permitting either steady-state protrusion or so-called retrograde flow of the polymer actin inward from the cell boundary. Both protrusion and retrograde flow can occur simultaneously. Their proportion to each other being inconstant, these two processes underlie the extremely complex, boiling appearance of the typical active cell surface. Whether under the conditions of steady-state protrusion or retrograde flow, the depth of the actin cortex may be unchanging, giving rise to the concept of "array treadmilling" of cortical actin, which emphasizes balanced polymerization and depolymerization that take place, respectively, at the outer edge and more internally. An integral process in the actin array treadmilling is termination of growth of the filament tips that are not in contact with the plasma membrane by capping protein. Polymerization is thereby restricted to the outer boundary of the cortex.

The aspect of this complex self-assembly process that makes it paradigmatic for the problems of cell morphogenesis is that the local interactions of the molecules involved do not define the global structure of the array directly. In particular, they do not, as such, define the orientation of the filaments with respect to the outer boundary, whose functional importance for the mechanics of the overall cell shape change was discussed above. At the same time, dendritic nucleation means that the orientation of the new filaments depends on the orientation of the existing ones. This circumstance sets the stage for Darwinian dynamics on the intracellular scale. As the actin filaments at the cell boundary proliferate by means of dendritic nucleation, the determination of the angle between the daughter and mother filaments by the Arp2/3 complex endows the filaments with a specific type of heredity of their orientation. Investigating the quantitative laws of this heredity and the mechanistic basis for the remaining conditions for Darwinian dynamics requires formalization of the spatial kinetics of branched actin polymerization.

Additional terminology and notation for quantitative conceptualization of the actin filament dynamics (Maly and Borisy 2001) begin with a description of the actin filament ends. Actin filaments are intrinsically polar, one end (called "barbed end") being favored for growth and the other end (called "pointed"), favored for shortening. The angle between a filament and its branch—branching angle—is concentrated around the mean $\psi = 67°$ with standard deviation $\sigma = 12°$ (Svitkina and Borisy 1999). The inheritance of orientation, therefore, is accompanied by both a systematic shift and mutation. Reproduction, heredity, and mutation having set the stage for evolution by natural selection, the exact trait (a function of the heritable property) on which the selection operates determines the evolution outcome. This consideration necessitates a focus on the spatial determinants of the kinetics of capping, which, as we shall see, is a critical determinant of the filament's proliferation potential.

Consider a filament nucleated at angle ϕ with respect to the normal to the leading edge (Fig. 5.1). When the plasma membrane advances at rate v and the said filament is propelling the protrusion, it must elongate, by addition of actin monomers to its barbed end, at a rate $s = v/\cos\phi$. Note that while we refer to v as the rate of the cell boundary protrusion, the formalism is the same if the boundary remains stationary and v represents the rate of cortical flow off the plasma membrane inward into the cell—or, indeed, if v represents the sum of the two rates. Having characterized the filament growth rate geometrically, we may note that, at the same time, the rate of filament elongation can be expressed kinetically as $s = \delta k_a M_a p$. In this expression, we have neglected the small rate of monomer dissociation at the growing tip and introduced δ, the elongation increment that results from addition of a monomer actin molecule. Further, k_a is the polymerization rate constant, M_a, the concentration of monomer, and p, the probability that the filament tip is not obstructed by the membrane but open due to thermal fluctuations (Mogilner and Oster 1996).

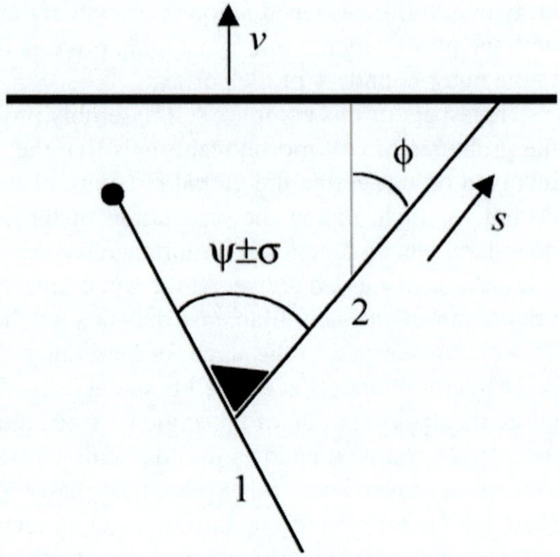

Fig. 5.1 Organizational unit of cortical actin network. *1*, mother filament, *2*, daughter filament, *heavy line*, plasma membrane, *solid lines*, filaments, *triangle*, Arp2/3 complex, *filled circle*, capping protein. Other notation is explained in the text. (Reproduced from Maly and Borisy 2001 with permission. Copyright 2001 National Academy of Sciences)

For a filament propelling the leading edge, $p = p_0/\cos\phi$, where $p_0 = v/\delta k_a M_a$ is the probability that a tip orthogonal to the leading edge is open, which is equivalent to the ratio of the rate of protrusion to the rate of elongation of a filament not in contact with the surface. Probability p depends on filament orientation and reaches unity when $|\phi|$ increases to a critical angle, $\theta = \arccos p_0$. For $|\phi|$ greater than the critical angle, polymerization does not keep up with protrusion, and the filament tip loses contact with the advancing membrane. Since nucleation of a filament depends on the Arp2/3 complex being activated by a WASP family member protein localized at the active sites of the plasma membrane, filaments oriented so that $|\phi| > \theta$ do not generate branches and are excluded from reproduction. Termination of filament growth similarly depends on orientation. For $|\phi| > \theta$, filament tips are not obstructed by the membrane and become terminated by association of capping protein at a rate, c, that is simply the normal biochemical rate of capping a free filament barbed end. For $|\phi| < \theta$, termination is possible only when the tip is open, and occurs at a rate $cp = cp_0/\cos\phi$. Hence, the smaller $|\phi|$, the longer the filament tip is not capped and the longer it stays at the leading edge and branches. In this way, the reproductive potential of a filament depends on its orientation, making orientation, in Darwinian terms, a selectable trait.

Dendritic nucleation leads to patterns of actin filament reproduction which depend on the relation between the angles ψ and θ. Although the analysis can be generalized, it is worth focusing on the form it takes for the values of ψ that are known in the systems assessed experimentally, namely, 60–90°. Also, branching of filaments constrained in the flat lamellipodium—an example of an active cortical site that is frequently seen in metazoan cell locomotion—is simpler than the fully three-dimensional case. It will be considered first. In this instance, and neglecting mutations for the time being ($\sigma = 0$), filaments with orientation ϕ can, *in potentia*, be nucleated on filaments with orientation $\phi \pm \psi$. Taking into account the restriction

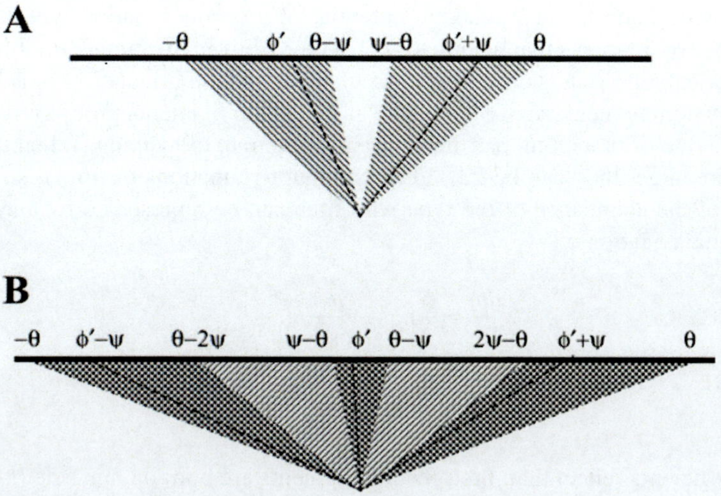

Fig. 5.2 Reproductive patterns of cortical actin filaments. (**a**) $\psi/2 < \theta < \psi$, (**b**) $\theta > \psi$. *Heavy line*, plasma membrane, *dashed lines*, reproductively connected filament angles, *hatched area*, range of possible heritable orientations, *cross-hatched area*, ranges populated by three-orientation reproductive types. ϕ' is a sample filament angle. $\psi = 67°$. (Reproduced from Maly and Borisy 2001 with permission. Copyright 2001 National Academy of Sciences)

of the reproductive range to $\pm\theta$, however, we can note more complex but also more prescriptive conditions for how the filament angles are inherited from generation to generation (Fig. 5.2). If $\theta < \psi/2$, the branching angle is greater than the acceptable range of angles, 2θ, for filaments to maintain contact with the membrane. Consequently, a branch of any existing filament would be oriented beyond the critical angle and reproduction could not be sustained. If $\psi/2 < \theta < \psi$, filament branches can remain within the critical angle and therefore reproduction can occur. However, not all filament orientations result in successful progeny. Filaments in the range θ–ψ to ψ–θ will generate branches beyond the critical angle, and thus their orientations will not be sustained in subsequent generations. Filaments in the range $-\theta \leq \phi \leq \theta-\psi$ may generate branches at $\phi-\psi$ or $\phi+\psi$, but only branches to one side, namely $\phi+\psi$, give rise to filaments which are in the acceptable range for sustained reproduction, $\psi-\theta \leq \phi \leq \theta$. Similarly, filaments in the range $\psi-\theta \leq \phi \leq \theta$ give rise to fecund branches in the range $-\theta \leq \phi \leq \theta-\psi$. Thus, within this regime ($\psi/2 < \theta < \psi$), filaments reproduce in two-generation cycles of orientations. If, however, $\theta > \psi$, then some filaments can nucleate a fecund branch to either side, and therefore sequences of three orientations in successive generations also become possible—for example, $(\phi-\psi, \phi, \phi+\psi)$, $|\phi| < \theta-\psi$. Sequences of four or more orientations cannot exist for $\psi > 60°$, because this would require at least one branch to be oriented at over 90° with respect to the normal to the surface. Thus, reproductive patterns for actual values of ψ occur as pairs or triplets of filament orientations. A given pair or triplet of orientations, which arise in the course of reproduction of filaments, has been termed a "type" (Maly and Borisy 2001). Since filaments can only reproduce within a type, reproductive potential is a property of a type, not of an individual filament.

Let us evaluate the reproductive potential of a two-orientation type first. The abundance of filaments that belong to such a type is given by number n_1 of filaments whose orientation is ϕ and number n_2 of filaments whose orientation $\phi+\psi$ is reproductively complementary to ϕ. The rate at which the filaments proliferate depends on *b,* the rate of branching per filament and *c,* the rate of capping. When the numbers *n* are large, they can be treated as continuous functions of time *t,* so that the change of the abundance of the type with time can be represented by a system of differential equations:

$$\begin{aligned}\frac{dn_1}{dt} &= \frac{b}{2}n_2 - \frac{cp_0}{\cos\phi}n_1,\\ \frac{dn_2}{dt} &= \frac{b}{2}n_1 - \frac{cp_0}{\cos(\phi+\psi)}n_2.\end{aligned} \tag{5.1}$$

The coefficients reflect that, first, fecund filaments are born on one side (*i.e.*, at the rate $b/2$) on filaments of the complementary orientation and, second, that filaments of orientation ϕ are capped at the rate $cp_0/\cos\phi$. The solution to this system of equations is

$$\begin{aligned}n_1(t) &= C_1\exp(\lambda_1 t) + C_2\exp(\lambda_2 t),\\ n_2(t) &= C_3\exp(\lambda_1 t) + C_4\exp(\lambda_2 t),\end{aligned} \tag{5.2}$$

where C_1, ..., C_4 are constants that depend on the initial conditions and λ_1, λ_2 are functions of the kinetic coefficients in Eq. (5.1). Of the two exponential terms, the one with the larger λ will describe the proliferation in the long run, dominating over the other term. Therefore, as long as branching prevails over capping, which is the basic condition for continuous cortical actin assembly, Eq. (5.2) reflects an asymptotically exponential increase in the abundance of the type. The rate of this asymptotic increase, *i.e.*, the larger λ, is a function of orientation ϕ that enters the kinetic coefficients in Eq. (5.1):

$$\begin{aligned}\lambda = f(\phi) = \frac{1}{2}cp_0\{&\sqrt{\left[\frac{\cos\varphi - \cos(\phi+\psi)}{\cos\varphi\cos(\phi+\psi)}\right]^2 + \left(\frac{b}{cp_0}\right)^2}\\ &- \frac{\cos\phi + \cos(\phi+\psi)}{\cos\phi\cos(\phi+\psi)}\}.\end{aligned} \tag{5.3}$$

Function $f(\phi)$ quantifies the reproductive potential of filaments whose orientation is ϕ. Obviously, the filaments of the complementary orientation $\phi + \psi$ have the same reproductive potential. Being the rate of exponential increase, $f(\phi)$ is by definition the Malthusian parameter of the type (Fisher 1999; Michod 1999). As introduced by R. Fisher, it "measures fitness to survive by the objective fact of representation in future generations," and can also be called Fisherian fitness (Michod 1999) of the type.

It can be emphasized that Eq. (5.3) was derived for the specific conditions for which Eq. (5.1) was written. More generally, the abundance of a type in the flat lamellipodium is given by numbers n_i of filaments with orientations ϕ_i, $i = 1, 2$, or 3, that belong to the type. After introducing χ which will be the largest of ϕ_i,

$$\chi = \begin{cases} \phi + \psi, \text{if } \phi \le \theta - \psi, \\ \phi, \text{if } \theta - \psi < \phi \le \theta, \end{cases}$$

we can arrange numbers n_i in the order of increase of ϕ_i to form a vector **n** and write, for large n_i and $\phi \ge 0$:

$$\dot{\mathbf{n}} = \mathbf{A}\mathbf{n}, \tag{5.4}$$

where the form of **A** depends on the relationship between the angles. Should it be that $|\psi - \theta| \le \chi < 2\psi - \theta$, we will have

$$\mathbf{A} = \begin{bmatrix} -cp_0 / \cos(\chi - \psi) & b/2 \\ b/2 & -cp_0 / \cos\chi \end{bmatrix}.$$

If, by contrast, $\chi \ge 2\psi - \theta$, then

$$\mathbf{A} = \begin{bmatrix} -cp_0 / \cos(\chi - 2\psi) & b/2 & 0 \\ b/2 & -cp_0 / \cos(\chi - \psi) & b/2 \\ 0 & b/2 & -cp_0 / \cos\chi \end{bmatrix}.$$

Fitness f is given by the largest eigenvalue of **A**. For $\phi < 0$, $f(\phi) = f(-\phi)$. Fitness is not defined for $|\phi| > \theta$, nor for any ϕ in the event that $p_0 > \cos(\psi/2)$, because filament reproduction in such situations cannot be sustained and therefore types do not exist.

Differentiation of f shows that for $\cos\psi < p_0 < \cos(\psi/2)$, it reaches maxima at $\phi = \pm\psi/2$. Therefore, the fittest under this condition is the two-generation type with symmetrically alternating orientations. Analogously, for $p_0 < \cos\psi$, the type consisting of three orientations $(0, \pm\psi)$ has the greatest fitness. Thus, optimal orientations exist at which reproductive potential is maximized.

Mutations are necessary to provide variation of orientation, on which natural selection can operate. Let the variation be Gaussian with root-mean-square deviation σ from the mean angle ψ between the two filaments. While the estimate of σ (12°) is known from experiments, the Gaussian form is consistent with the origin of the variation being thermal fluctuations that become trapped in the network as the filaments grow. Taking variation into account, number density $n(\phi, t)$ of filaments at the leading edge of the lamellipodium as a function of their orientation and time evolves according to the following equation:

$$\frac{dn(\phi,t)}{dt} = b\frac{1}{\sqrt{8\pi}\sigma}\int_{\theta}^{-\theta}\left(e^{-\frac{(\omega+\psi-\phi)^2}{2\sigma^2}} + e^{-\frac{(\omega-\psi-\phi)^2}{2\sigma^2}}\right)n(\omega,t)d\omega$$
$$-c\frac{p_0}{\cos\phi}n(\phi,t). \tag{5.5}$$

The kernel under the integral sign reflects the probability density that a filament at angle ϕ is nucleated on a filament at angle ω. The pseudo-first-order rate constants of branching and capping, b and c, should be regarded as unknown functions of N, the total number of filaments in contact with the leading edge. This accounts for the fact that the actin network in question is at steady state, leaving aside the kinetic form of the feedback from the polymer mass (recall that it is a co-polymer of actin with Arp2/3 and the capping protein) in the vicinity of the leading edge to the local specific rates of filament branching and capping. In general, density function $q(\phi, t)$ of the distribution of angle ϕ is defined by $n(\phi, t) = q(\phi, t)N(t)$. The Maly-Borisy formalism laid out here applies to any given steady state in which neither N nor $q(\phi)$ changes with time. Under such conditions, it follows from Eq. (5.5) that the steady-state density function $q_{ss}(\phi)$ is the eigenfunction, and the steady-state ratio of capping and branching specific rates, c_{ss}/b_{ss}, is the eigenvalue of what has been termed the branching-capping operator, as expressed in the following eigenproblem:

$$\frac{\cos\phi}{p_0\sqrt{8\pi}\sigma}\int_{\theta}^{-\theta}\left(e^{-\frac{(\omega+\psi-\phi)^2}{2\sigma^2}} + e^{-\frac{(\omega-\psi-\phi)^2}{2\sigma^2}}\right)q_{ss}(\omega)d\omega = \frac{c_{ss}}{b_{ss}}q_{ss}(\phi). \tag{5.6}$$

Analysis shows that for every p_0, there is a unique solution, in which c_{ss}/b_{ss} is of the order of unity and $q_{ss}(\phi)$ depends on p_0 as shown in Fig. 5.3. The maxima of $q_{ss}(\phi)$ correspond well to the maxima of $f(\phi)$. Thus, the fittest, as defined by the most rapidly proliferating, dominate in the steady-state population. According to the bifurcation diagram (Fig. 5.3a), the predominant orientations are close to 0 when p_0 is small, abruptly change to approximately ±35° for intermediate values of p_0, and gradually become small again as p_0 approaches 1.

The relationship between the fitness and steady-state density functions is illuminating of the intracellular evolutionary origin of the cortical actin self-organization. The filament types differ in their capacity to proliferate as expressed by Eq. (5.1) and $f(\phi)$ in terms of the underlying kinetics and geometry. The types transmute because of the inter-generational dispersion (mutation) of filament orientation, while the number of filaments remains constant, as expressed in Eq. (5.6), whose solution is $q_{ss}(\phi)$. Taking this into account, the correspondence of the maxima of the fitness function $f(\phi)$ to those of the steady-state density function $q_{ss}(\phi)$ highlights the action of Darwinian selection of actin filaments in the lamellipodium.

Comparing $q_{ss}(\phi)$ and $f(\phi)$ in more detail, it should be noted that in the three-orientation type, the middle orientation is twice as abundant as each of the two others. In combination with the rather strong mutation (σ), this leads to there being only

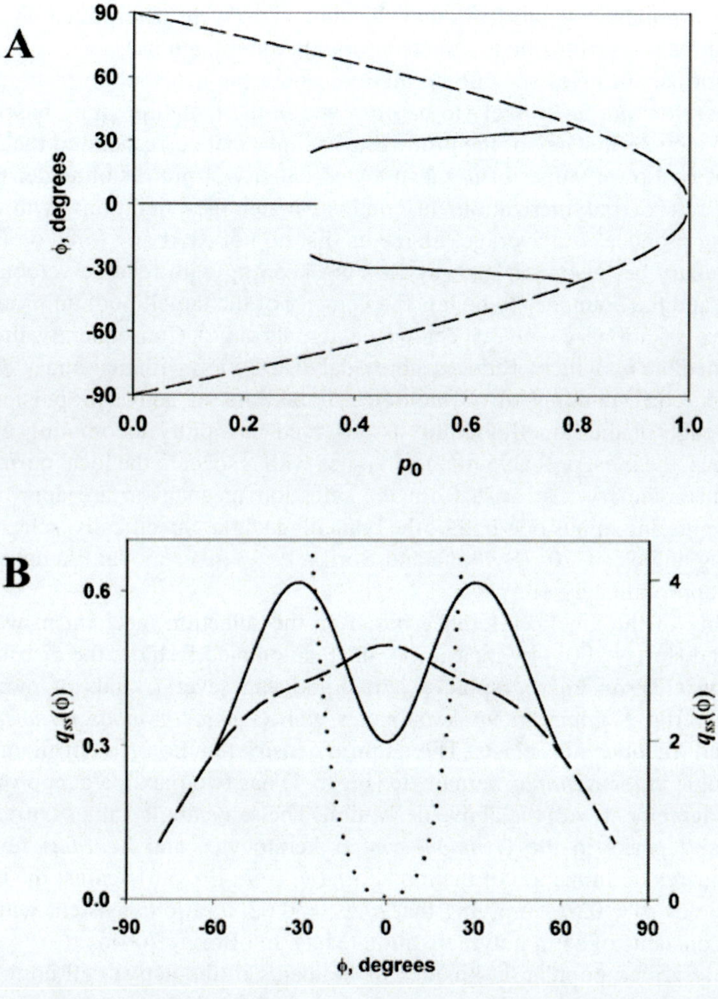

Fig. 5.3 Steady-state orientational distribution of actin filaments in a lamellipodium. (**a**) Modal angle values (*solid lines*) and distribution support boundaries (*dashed lines*). The modal angles coincide with the boundaries where not plotted. (**b**) Angular density function at $p_0 = 0.2$ (*dashed curve, left axis*), $p_0 = 0.5$ (*solid curve, left axis*), and $p_0 = 0.9$ (*dotted curve, right axis*). $\psi = 67°$, $\sigma = 12°$. (Reproduced from Maly and Borisy 2001 with permission. Copyright 2001 National Academy of Sciences)

one maximum in $q_{ss}(\phi)$ as compared to three in $f(\phi)$, when $p_0 < \cos\psi$. The strong effect of mutation is further seen in the regime that corresponds to $p_0 > \cos(\psi/2)$, where, although pure types do not reproduce themselves (and so $f(\phi)$ is not defined), the assembly of the network can nevertheless persist ($q_{ss}(\phi) \neq 0$, Fig. 5.3) by relying entirely on errors in the orientation inheritance. For the same reason, under the conditions where $\cos\psi < p_0 < \cos(\psi/2)$, all angles $-\theta < \phi < \theta$ are represented, despite the fact that some types do not reproduce. Slight displacements of the maxima and bifurcation points of $q_{ss}(\phi)$ from their locations that were determined analytically for $f(\phi)$ are also attributable to effects of orientation mutations.

The formalism gives specific predictions about the distribution of the filament orientation that are most likely to be observed in the lamellipodium, based on the relative kinetic efficiency of the protrusion regimes that correspond to the different ranges of p_0. Lower values of p_0 mean lower velocity of protrusion under the same possible rate of polymerization. In contrast, under the conditions with high p_0, branching produces, on average, filaments that do not exert any force on the outer cell boundary because their growth does not keep up with relative velocity of the network and the boundary (which is the velocity of the lamellipodium protrusion if the cortex is stationary with respect to an external frame). Consequently, the regime of intermediate p_0, which produces a bimodal distribution with maxima at $\phi = \pm\psi/2$, is the one corresponding to physiological conditions of active propulsion of the leading edge of the lamellipodium, as observed in rapidly locomoting metazoan cells. Thus, the most probable filament angles with respect to the local normal to the outer cell boundary that arise from the selection mechanism are approximately equal to plus and minus one-half of the branching angle. Specifically, selection with branching angle $\psi \approx 70°$ (Svitkina and Borisy 1999) gives modal filament orientations of approximately ±35°.

Validity of the kinetic inference based on the selection mechanism was tested (Maly and Borisy 2001) by measuring, in an automated fashion, the distribution of angles on electron micrographs of lamellipodia in several kinds of motile cells including frog *Xenopus laevis* keratocytes, fish *Gymnocorymbus ternetzi* keratocytes, and *Xenopus* fibroblasts. The empiric density function $r(\phi)$ of filament incidence angle in the *Xenopus* keratocyte (Fig. 5.4) has two maxima at approximately ±35° in agreement with the above derivation. The same angle values correspond to the density peaks in the *Gymnocorymbus* keratocytes and *Xenopus* fibroblasts. Least-squares optimization of the model ($q_{ss}(\phi)$ from Eq. 5.6) against the measurements yields $p_0 = 0.57$, implying that $c_{ss}/b_{ss} = 0.62$, a ratio consistent with known kinetic constants of actin polymerization (Maly and Borisy 2001).

The described population kinetics of filaments in the active cell cortex is also consistent with the elastic Brownian ratchet model for the process of force generation at the actin-membrane interface that was put forward by Mogilner and Oster (1996). The controlling parameter p_0, which is employed kinematically in the evolutionary theory, is, in dynamic terms, a Boltzmann factor—the exponential of the work, in units of thermal energy, that is done by polymerization of a filament to produce a filament-membrane relative displacement that arises from incorporation of one actin monomer. The value of p_0, therefore, depends on the mechanical load, *i.e.*, on the force that resists the displacement. Accordingly, this critical parameter in the selection mechanism approaches zero at high load, and becomes equal to unity in the limit where the resistance to membrane protrusion (or to retrograde flow) vanishes.

The spatial-kinetic theory of the actin cortex highlights how immediate natural selection of supramolecular complexes works on the results of expression of genetic

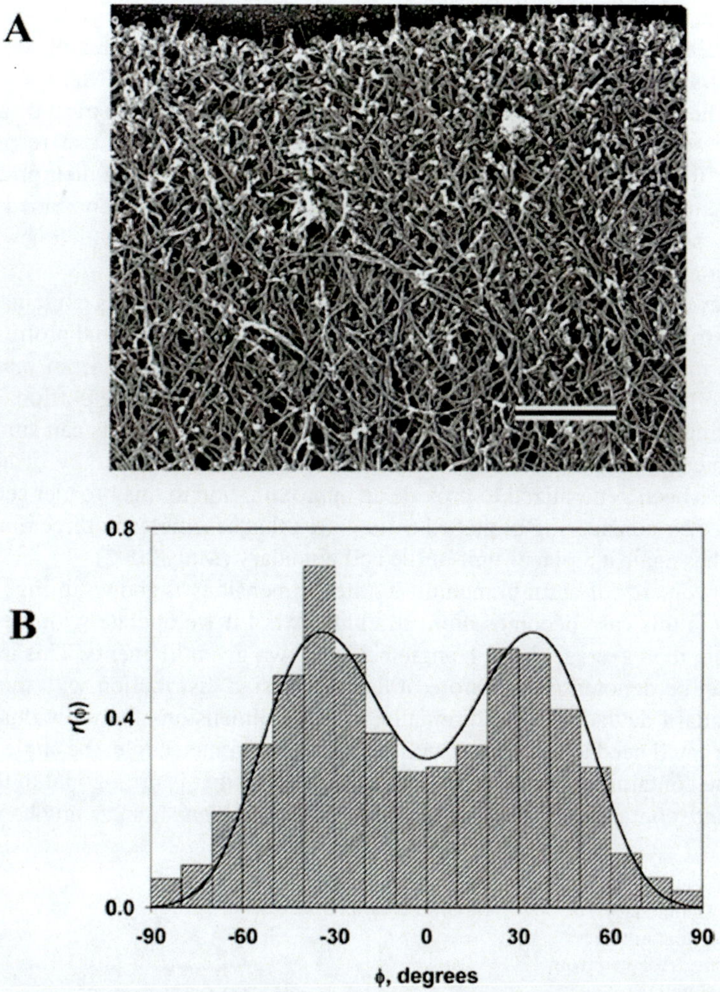

Fig. 5.4 Actin filaments in a lamellipodium. (**a**) Electron micrograph of the leading edge of a moving *Xenopus* keratocyte, with the plasma membrane removed. The cell edge is at the top of the image. *Scale bar*, 0.2µm. (**b**) Histogram of angles between filaments and the normal to the leading edge in a *Xenopus* keratocyte lamellipodium. *Curve*, theoretical approximation. (Reproduced from Maly and Borisy 2001 with permission. Copyright 2001 National Academy of Sciences)

information and self-assembly. The cortical actin example is important due to the central role of the cortex in cell and tissue morphogenesis. This example, however, is unlikely to be unique, given that other processes involved in the biogenesis of subcellular structures employ inexact self-assembly of new constituent units on pre-existing structural patterns. Such circumstances form the qualitative substrate on which Darwinian selection may arise when specific kinetic conditions are met. What makes this likely—considering that the kinetic conditions in question are themselves controlled by the long-term Darwinian evolution of cells—is that the intracellular evolutionary self-organization is economical in terms of the cellular genetic information required to specify the needed structure (Maly and Borisy 2001). The parsimony is particularly important when the structure must be adaptive. In the case of the processes governed by cortical actin, cells can respond, for instance, to chemotactic stimuli or substratum cues by reorienting their protrusions. The constitutive existence of a spectrum of filament orientations, combined with the inexactness of the branching angle, provides the material from which new favored orientations can be selected.

Surveyed above, the kinetic self-organization of actin filaments explains the filament distributions observed in very flat, essentially two-dimensional protrusive portions of motile metazoan cells. However, this geometry differs from many other motile cells, such as pollen tubes and root hairs, whose protrusive portions advance by forming a cylindrical tube. Active cell cortex in metazoan cells can similarly be represented by three-dimensional amoeboid protrusions. The theory of actin networks has been generalized to provide an approximation to this broader set of morphologies by considering branched actin cortex that assembles in three dimensions and pushes against a planar unit of the cell boundary (Maly 2002).

The geometry of actin branching in three dimensions is shown in Fig. 5.5. The analysis in this case becomes more straightforward if we operate with the individual, rather than average, branch angle between two given filaments. This individual angle can be denoted β. As before, it has a Gaussian distribution with the mean ψ and standard deviation σ. Additionally, in three dimensions the individual branch geometry will need to be defined with the rotational branch angle: the angle between the plane containing the two filaments and the plane that is orthogonal to the membrane and contains the mother filament. The rotational branch angle, which is

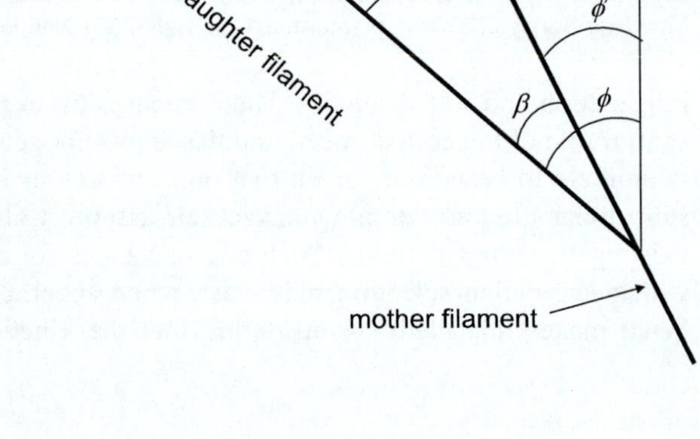

Fig. 5.5 Geometry of actin branching in three dimensions. (Adapted from Maly 2002 with permission)

distributed uniformly between 0 and 2π, can be denoted ρ. It can be verified that if the incidence angle of the mother filament on the propelled membrane is ϕ', then the incidence angle of the daughter filament is $\phi(\phi', \beta, \rho) = \arccos(\cos\rho\sin\phi'\sin\beta + \cos\phi'\cos\beta)$. Given the mother filament incidence angle, the probability density of the daughter filament incidence angle can be denoted $B(\phi, \phi')$. It can be obtained with the Monte Carlo method.

The rate of production of filaments whose incidence angle is ϕ on a filament whose incidence angle is ϕ' can be expressed as $bB(\phi, \phi')$, where b, as before, is the branching rate per filament. Then the following equation will be the analogue of Eq. (5.6) in the three-dimensional case:

$$\frac{\cos\phi}{p_0}\int_{\theta}^{0} B(\phi,\phi')q_{ss}(\phi')d\phi' = \frac{c_{ss}}{b_{ss}}q_{ss}(\phi).$$

The solution to this equation is plotted in Fig. 5.6. The theory predicts that branching in three dimensions results in an almost constant modal value of the incidence angle of about 45° at any protrusion rate. This is in contrast to branching in two dimensions, where there is bifurcation at $p_0 \approx \cos\psi$ (cf. Fig. 5.3).

Notably, the equation for the steady-state distribution of the incidence angle of propulsive actin filaments can be rewritten in the form

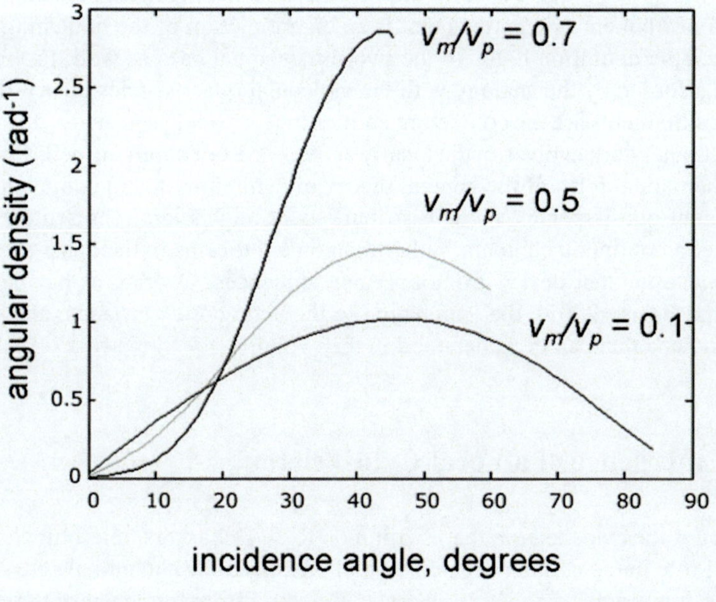

Fig. 5.6 Steady-state distribution of filament incidence angles with the cell membrane in three dimensions. Ratio v_m/v_p of the membrane velocity to the velocity of unrestricted polymerizing barbed end is equivalent to parameter p_0 in the text. $\psi = 70°$, $\sigma = 10°$. (Reproduced from Maly 2002)

$$\int_{\theta}^{0} \tilde{B}(\phi,\phi')q_{ss}(\phi')d\phi' = c_{ss}q_{ss}(\phi), \quad \tilde{B}(\phi,\phi') = \frac{b_{ss}}{p_0}\cos\phi B(\phi,\phi'),$$

$$c_{ss} = \frac{\int_0^\theta \int_0^\theta \tilde{B}(\phi,\phi')q_{ss}(\phi')d\phi' d\phi}{\int_0^\theta q_{ss}(\phi)d\phi},$$

which is the continuous analogue of the molecular quasi-species equation:

$$\mathbf{Wx} = \lambda\mathbf{x}, \quad \lambda = \frac{\sum \mathbf{Wx}}{\sum \mathbf{x}},$$

where x_i is the concentration of genetic sequence i and W_{ij} is the rate constant of production of sequence i by sequence j (Eigen et al. 1989; Nowak 1992). This transformation reveals that the formalism for the selection of actin filaments conforms with the universal formalism of Darwinian evolutionary dynamics as outlined by Page and Nowak (2002).

Summarizing, the three-dimensional generalization of the evolutionary theory of self-organization of a propulsive actin network is useful for studies of architecture of such cell morphologies as root hairs, pollen tubes and amoeboid pseudopodia. In the generalized theory, filaments of different orientations are treated as chemical species whose reproduction is coupled through branching with the variable branch angle and branch rotation angle. The isomorphism of the equation for the selected distribution of filament orientations with the molecular quasi-species equation leads to the definition of a quasi-species of propulsive branching actin filaments as a distribution of filament orientations that is an eigenfunction of the branching-capping operator. This definition holds in the two-dimensional case as well. From the perspective afforded by the analogy with the molecular quasi-species theory, the quasi-species of filaments are the competing entities that are being selected as the structure of the actin network evolves in the locally active cortex or a moving cell. Concerning the mathematical form of the general theory of evolution, it is of some interest that the filament quasi-species, being distributions of angles, are represented by functions over a continuous domain, rather than on a lattice as in the quasi-species theory applications that deal with biopolymer sequences. Overall, as has been noted (Maly and Borisy 2001), the application to the actin cortex broadens the spectrum of phenomena that can be understood in the evolutionary framework.

5.2 Emergence of a Focal Cytoskeleton

As a part of the cytoskeleton that is common to all Eukaryota, microtubules play an essential role in the cellular organization of advanced life. In animal cells, microtubules are commonly arranged in an aster, stretching from the center of the cell to the periphery. Typically, this organization results from initiation of microtubule assembly at the centrosome, a specialized aggregate positioned near the cell center and representing the dynamic center of the cell (Boveri 1887). The centrosome's function is mediated by nucleation of polymerization of the soluble cytoplasmic protein tubulin into multi-stranded polymer forms—microtubules. The centrosome also stabilizes the so-called minus ends of the microtubules, which stay attached to it after the nucleation. The opposite "plus" ends display dynamic instability of polymerization, *i.e.*, alternating phases of elongation and shortening. Using the resulting steady-state arrangement, in which the microtubules radiate from the cell center with the minus ends in the center and plus ends oriented toward the periphery, molecular motors that recognize the microtubule polarity ferry their cargo selectively to or from the cell center (Bray 2001). Through its transport-directing functionality, the centrosome-based polar structure of the microtubule array confers the centered dynamical organization to the distribution of other organelles and the cell as a whole.

The radial microtubule array can also form spontaneously in the absence of the centrosome by means of a process that is regarded as a paradigm of self-organization of macromolecular ensembles into the functional systems of living cells (Sharp et al. 2000; Kirschner et al. 2000). Initially, the process was thought to involve physical arrangement of microtubules into the clustered, radial pattern by molecular motors, a process observed in cell extracts and biochemical reconstructions (Verde et al. 1991; Nedelec et al. 1997). Understanding of the centered array self-organization in vivo was advanced by studies on centrosome-free cell fragments, which highlighted the fact that during the spontaneous array formation in the intact, dense cytoplasm, individual microtubules remain stationary, *i.e.*, the process does not involve any physical transport of the microtubules (Vorobjev et al. 2001; Malikov et al. 2005). At the same time, the formation of the radial array is dependent on microtubule assembly and disassembly (Rodionov and Borisy 1997a), as was noticed also in the foundational studies on long and narrow cell fragments that similarly self-organize central symmetry in their microtubule systems (McNiven et al. 1984; McNiven and Porter 1986; McNiven and Porter 1988). Remarkably, the absence of the centrosome alters the pattern of microtubule dynamics relative to the already described picture: The minus ends that are not anchored depolymerize and release tubulin, and this flux of tubulin into the soluble dimer pool drives a persistent growth of the free plus ends. This "treadmilling" polymerization pattern (Rodionov and Borisy 1997b; Vorobjev et al. 1999; Rodionov et al. 1999) is similar to the one encountered in the last section in connection with actin polymerization kinetics. As with the intrinsically non-equilibrium polymerization of actin *in vivo*, treadmilling of microtubules is possible because hydrolysis of GTP bound to tubulin subunits in the course of polymerization makes microtubule assembly a thermodynamically directional process coupled to the metabolism of the living cell. As the acentrosomal radial array is forming, new microtubules polymerize from the pool of soluble tubulin, while the pre-existing ones are depolymerizing completely. Furthermore, the assembly of new microtubules in the acentrosomal cell fragments is nucleated by material that is transported toward the minus ends of existing ones. Once this material, driven by molecular motor dynein, is accumulated in the center of the cytoplasmic domain, the system's form and dynamics become equivalent to those in cells that inherit pre-formed centrosomes through cell division.

These observations spurred the development of the quantitative theory of the cytoplasm self-centering process and led to the recognition that the self-organization mechanism operating in the absence of the centrosome must contribute also to the integrity of a centrosome-based microtubule array as well as serve as its evolutionary precursor (Maly and Borisy 2002; Maly 2002; Cytrynbaum et al. 2004; Malikov et al. 2005; Cytrynbaum et al. 2006). In particular, it has been mathematically demonstrated (Maly and Borisy 2002) how microtubule treadmilling dynamics in the confines of the cytoplasmic domain boundary results in an establishment of a degree of radial polarity in the microtubule distribution which can support the directional redistribution of microtubule-nucleating material toward the center. Analysis of the kinetics involving both treadmilling and the nucleating material transfer, using a coupled partial differential equations formalism, showed that this minimal system is indeed capable of displaying the emergence of a tight radial array centered in the cytoplasm (Maly 2002). The cited subsequent work elaborated on these results both computationally and experimentally, leading to validated discrete numerical simulations of the process.

The Maly-Borisy theory sharpens the question about the emergence of the focal cytoskeleton as follows: what drives the microtubule-nucleating material to the center of the cytoplasmic domain, if the initial distribution of microtubules is random? The dynein molecular motors, which are responsible for the transport of the nucleating material, do not have any means of selecting the microtubules with the correct orientation. Therefore, spontaneous alignment of the microtubules themselves in the bounded cytoplasmic domain is the fundamental process giving direction to further self-organization. Since the microtubules in question are physically immobile, the organizing force must derive from the thermodynamically directional microtubule polymerization dynamics—nucleation, assembly, and disassembly—and develop in reference to the cellular boundary as the only preexisting spatial marker in the system.

As a treadmilling microtubule grows at its plus end and shortens at its minus end, it effectively moves through the cytoplasm with its plus end forward, even though all tubulin subunits within it are stationary, as established by the observations. Unless the microtubule is aligned with the center of the cytoplasmic domain by chance, the said apparent translocation alters the microtubule's orientation vis-à-vis the center. Maly and Borisy (2002) derived the steady-state microtubule distribution that results from the treadmilling dynamics in a bounded domain of cytoplasm with random nucleation (see also Maly 2002).

Consider a flat, round domain of cytoplasm, as in the experiments with cell fragments (Rodionov and Borisy 1997a; Vorobjev et al. 2001; Malikov et al. 2005). It will be treated as a circle, and microtubules as straight segments. Their ends displace at constant rates along the axes of the segments as a result of microtubule assembly at one end and disassembly at the other end. New microtubules appear by spontaneous nucleation in the cytoplasm (Vorobjev et al. 1997; Rodionov et al. 1999). The cytoplasm, apart from the microtubules, is considered to be isotropic and homogeneous. As a consequence, random orientation of newly nucleated microtubules and a constant rate of nucleation per unit area can be further assumed. For the time being, we will treat all minus ends as depolymerizing. When the minus end is shortening all the time after the nucleation of a given microtubule, microtubule formation occurs as a result of the rate of the plus end elongation being higher than the rate of the minus end shortening. Such an inequality of the rates exhibited by free plus and minus ends is characteristic of the kinetic conditions of microtubule polymerization observed during the radial array formation (Vorobjev et al. 2001). However, when a microtubule plus end arrives at the cell boundary, its growth stops. The shortening from the minus end continues, resulting eventually in a complete disassembly of the microtubule (Rodionov and Borisy 1997a).

The tubulin polymerization in the experimental cytoplasmic domains appears to be at steady state, and such a condition is compatible with the described kinetics of the ends. At any given time, the number of shortening ends exceeds the number of elongating ends, because all minus ends are shortening but some plus ends are paused at the boundary. The net balance of polymerization and depolymerization can be maintained, since the rate of elongation of an individual plus end exceeds the rate of shortening of an individual minus end. Any given microtubule, however, slowly grows in length as it "treadmills" across the cytoplasm, and then releases all of its polymerized tubulin after the plus end reaches the boundary.

Let us now consider an arbitrary chord AB crossing the cytoplasmic domain and introduce coordinate axis x, on which the chord AB lies, in such a way that the coordinate of point A is $x = 0$, and the coordinate of point B is $x = L > 0$. We shall consider microtubules that geometrically belong to the chord AB and whose direction, from the minus ends to the plus ends, coincides with the direction of the x axis. Such microtubules will be called AB microtubules. We will also say that a microtubule is in state 1, if its plus end is growing, and that it is in state 2, if its plus end is paused at the domain boundary. Let $m_1(t, x_m, x_p)$ be the concentration, at time t, of AB microtubules in state 1, whose minus ends are located at $x = x_m$ and whose plus ends are at $x = x_p$. Similarly, let $m_2(t, x_m)$ be the concentration, at time t, of AB microtubules in state 2, such that their minus ends are at $x = x_m$. The rate of plus end growth will be denoted v_p, the rate of minus end shortening, v_m, and the rate of microtubule nucleation per unit volume of the cytoplasmic domain, n. Then the following equations describe the dynamics of the AB microtubules:

$$\frac{\partial}{\partial t} m_1\left(t, x_m, x_p\right) = -v_p \frac{\partial}{\partial x_p} m_1\left(t, x_m, x_p\right) - v_m \frac{\partial}{\partial x_m} m_1\left(t, x_m, x_p\right);$$

$$\frac{\partial}{\partial t} m_2\left(t, x_m\right) = v_p m_1\left(t, x_m, L\right) - v_m \frac{\partial}{\partial x_m} m_2\left(t, x_m\right);$$

$$m_1\left(t, x_m, x_p\right) = \frac{n}{2\pi\left(v_p - v_m\right)}, \quad x_m = x_p;$$

$$m_1\left(t, x_m, x_p\right) = 0, \quad x_m = 0, \quad x_p > 0; \quad m_2\left(t, 0\right) = 0. \tag{5.7}$$

Dropping the time argument in the functions m, one can write the time-independent solution to Eq. (5.7) that describes the steady-state distribution of the AB microtubules as

$$m_1(x_m,x_p)=0,\quad x_m<\frac{v_m}{v_p}x_p;$$
$$m_1(x_m,x_p)=\frac{n}{2\pi(v_p-v_m)},\quad \frac{v_m}{v_p}x_p\le x_m\le x_p;$$
$$m_2(x_m)=0,\quad x_m<\frac{v_m}{v_p}L;$$
$$m_2(x_m)=\frac{n}{2\pi(v_p-v_m)}\left(\frac{v_p}{v_m}x_m-L\right),\quad x_m\ge\frac{v_m}{v_p}L. \tag{5.8}$$

The steady-state concentration of minus ends of AB microtubules at the position with coordinate x is

$$p(x)=\int_L^x m_1(x,x_p)\,dx_p+m_2(x)=\frac{n}{2\pi v_m}x. \tag{5.9}$$

With this in mind, the steady-state concentration of AB microtubules crossing the position with coordinate x is

$$q(x)=\int_L^x\int_x^0 m_1(x_m,x_p)\,dx_m dx_p+\int_x^0 m_2(x_m)\,dx_m=bx^2,$$
$$b\equiv\frac{(v_p-v_m)n}{4\pi v_p v_m}. \tag{5.10}$$

The analysis derived up to this point applied to an arbitrarily shaped domain of cytoplasm, provided that the dimensions of n were in accordance with the dimensionality of the domain. Let us now specifically consider a flat round domain of cytoplasm, and select a point, M, that belongs to chord AB and is located at distance r from the center of such a domain. Further, let ϕ be the angle of the x axis with respect to the radius of the domain that is drawn through point M. If the radius has length R, then the x coordinate of point M is

$$x(r,\phi)=\sqrt{R^2-(r\sin\phi)^2}+r\cos\phi. \tag{5.11}$$

Using Eqs. (5.9) and (5.11), we can express the total steady-state concentration of minus ends (not only those of AB microtubules) at any point located at distance r from the center of the cytoplasmic domain as follows:

$$u(r)=\int_{2\pi}^0 p(x(r,\phi))\,d\phi=\frac{2nR}{\pi v_m}E\left(\frac{r}{R}\right), \tag{5.12}$$

where E is the complete elliptic integral of the second kind (Abramowitz and Stegun 1972). By substituting Eq. (5.11) into (5.10), we derive

$$f(r,\phi) = b\left(\sqrt{R^2 - (r\sin\phi)^2} + r\cos\phi\right)^2. \tag{5.13}$$

This function, f, gives the concentration of microtubules that pass through a certain point at distance r from the center of the flat, round cytoplasmic domain in the direction, from minus to plus ends, that differs from the radial direction by angle ϕ. On integration over the angle,

$$\int_{2\pi}^{0} f(r,\phi)\,d\phi = \frac{(v_p - v_m)nR^2}{2v_p v_m}, \tag{5.14}$$

we obtain that the steady-state concentration of polymer tubulin is independent of position within the cytoplasmic domain.

The steady-state distributions of orientations of the randomly nucleated and treadmilling microtubules are shown in Fig. 5.7 for different distances from the center of a flat, round cytoplasmic domain. (The microtubule concentration is presented in accordance with Eq. (5.13), as f in the units of bR^2.) At any distance from the center, the distribution has a peak indicating that radial microtubules oriented with their minus ends toward the center predominate in the steady-state array. Close to the center, the distribution of the microtubule orientations approaches a uniform one because of the radial symmetry of the round cytoplasmic domain. Away from the center, the distribution becomes narrowly concentrated about the radial orientation with the minus ends toward the center. Thus, treadmilling of randomly nucleated microtubules leads spontaneously to a polar arrangement that is structurally reminiscent of the centrosome-based organization and possesses the same basic functional potential, namely, the ability to direct transport to and from the center of the bounded domain of cytoplasm, depending on the polarity of the microtubule-associated molecular motors. Although, as noted, the orientational distribution in the center is in both instances uniform with respect to the angle, the distribution arising from treadmilling lacks the defined focal point where the minus ends and polymer tubulin overall are aggregated. It, therefore, would not be able to support the typical "packed" molecular motor-driven assembly of membranous organelles near the cell center with the same efficiency.

The mixed organization of microtubules in the center of the centrosome-free cell fragment and more complete polarity toward the domain boundary was noticed already in the pioneering work of Porter and colleagues (McNiven et al. 1984; McNiven and Porter 1986; McNiven and Porter 1988), which also underscored the importance of tubulin polymerization dynamics in its emergence. The statistically polarized array's ability to direct transport of material toward the center of the fragment led to the formulation of the concept that microtubules determine the direction of net transport by the principle of "majority rule" (McNiven et al. 1984; McNiven and Porter 1986).

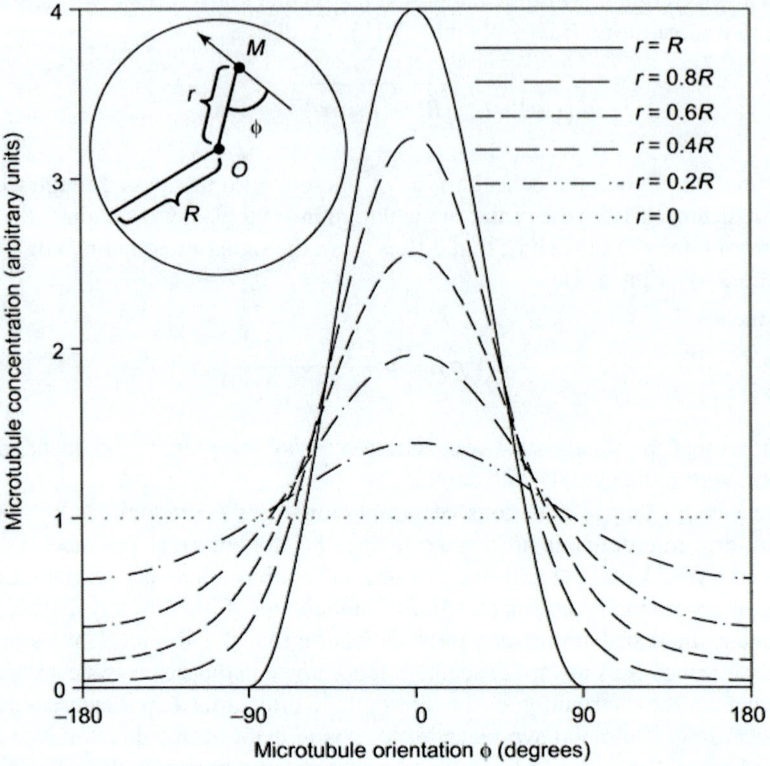

Fig. 5.7 Steady-state concentration of treadmilling microtubules of various orientations in a flat, round cytoplasmic domain. Inset: Schematic of a microtubule (*arrow*) in the round domain, with notation. *Arrowhead*, plus end. Reproduced from Maly and Borisy (2002). Reprinted from Trends in Cell Biology, Vol. 12, Ivan V. Maly, Gary G. Borisy, Self-organization of treadmilling microtubules into a polar array, 462–465, copyright 2002, with permission from Elsevier

The characteristic time for the visible establishment of a radial array in cell fragments is approximately 10 min (Rodionov and Borisy 1997b). In the treadmilling mechanism, the self-organized steady state is reached in the time necessary for a disassembling minus end to cross the diameter of the cytoplasmic fragment. At that time, no initially present microtubule remains in the cytoplasm, and the distribution of the microtubules becomes independent of the initial distribution and, therefore, time-independent. A typical melanophore fragment is 20μm across, and the mean apparent velocity of the minus end is 5.1μm/min (Vorobjev et al. 2001), setting the duration of the treadmilling-based microtubule rearrangement at approximately 4 min. Thus, the treadmilling-based rearrangement plays out in the beginning phase of the overall self-organization process.

Treadmilling of microtubules is sufficient to align the overall microtubule system polarity with the cell center, but it is not sufficient for organization of a true radial array with a tightly defined center. An additional mechanism, microtubule nucleation by the intracellular material driven by molecular motor dynein, and likely by dynein itself, has been identified as contributing to the overall self-organization process (Vorobjev et al. 2001). In contrast to treadmilling, dynein-facilitated nucleation of microtubules operates without reference to the boundary of the cytoplasmic domain and thus cannot by itself localize to the domain's center. However, as noted by Maly and Borisy (2002), the combination of treadmilling with motor-facilitated nucleation provides the means for establishment of the tight radial array. After any degree of the statistical center-boundary polarity has been established in the microtubule system due to treadmilling, the microtubule-nucleating material that is transported toward minus ends becomes directed to the center of the cytoplasmic domain. Reflecting this central accumulation of the nucleating material, newly appearing minus ends tend to originate in the center, progressively defining a radial organization in the system. The kinematics of the molecular motor-driven transport through a system of cytoskeletal fibers have consequences for the overall kinetics of the nucleating material redistribution simultaneously with microtubule treadmilling. However, the course of the compound process is predefined by the basic polarity development via treadmilling. These considerations led to the recognition of the array that is self-organized by treadmilling as the primordial element of the eukaryotic cytoplasmic order (Maly and Borisy 2002).

Polarity refinement by transport of microtubule nucleators has been analyzed on the basis of a simple formalism for simultaneous microtubule treadmilling and transport of microtubule nucleators in one dimension (Maly 2002). The one-dimensional treatment is not only sufficient conceptually, but also directly applicable to the cited experiments of the Porter group, which utilized long narrow cell fragments. In the context set by the experiments on melanophore cells (*e.g.*, McNiven et al. 1984; Vorobjev et al. 2001), which are paradigmatic for this line of experimental exploration, the term "nucleator" was adopted for the complex of a melanosome and dynein. This complex is both motile on existing microtubules (toward the minus end) and capable of nucleating polymerization of new microtubules. Once the nucleator is engaged with a microtubule, it moves to the corresponding minus end at a constant velocity v. Upon reaching the minus end, it disengages from the microtubule and engages with another microtubule, which is chosen at random among the locally available microtubules, irrespective of their polarity. Microtubules are nucleated in a random direction at a rate that is proportional to the local concentration of the nucleators, with the proportionality factor n.

In the combined treatment with nucleator dynamics, we can adopt a somewhat different description of microtubule treadmilling than earlier in this chapter. Specifically, we will assume that upon nucleation, the plus end of a microtubule elongates at a constant speed taken for the unit speed, while the minus end is stable. Destabilization of the minus ends has first-order kinetics with rate constant r. Upon destabilization, the minus end disassembles at a rate that equals the rate of assembly of polymerizing plus ends. Thus, the formalism with nucleators involves "pure" treadmilling, whereby a microtubule whose ends are both dynamic does not change length. Irrespectively, plus ends that arrive at the boundary of the cytoplasm stop growing, and the minus end disassembly proceeds until the microtubule disaggregates completely.

This kinetic picture can be formalized by numbering the dynamic states of microtubules. State 1 has a stable minus end and a growing plus end. State 2 has a stable minus end and a plus end arrested at the boundary. State 3 has a disassembling minus end and a growing plus end. State 4 has a disassembling minus end and a plus end arrested at the boundary. Figure 5.8a diagrams the possible transitions between the states. Further, the orientation of a microtubule whose plus end points in the direction of the coordinate axis can be given number 1, and the opposite orientation, number 2. The dynamic type of a microtubule can then be fully described by an index, *ij*, with *i* being the orientation and *j*, the dynamic state. We will denote the coordinate of the minus end as x and the coordinate of the plus end as y, taking the length of the cytoplasmic domain for the unit length. Figure 5.8b shows two sample trajectories tracing the life history of a microtubule represented by a point in the (x, y) space.

To proceed to derivation of microtubule and nucleator distributions, we need, first, to introduce the number density of microtubules. Its dimensionality will depend on the microtubule's dynamic state. The microtubule number density can be denoted as $m_{ij}(t, x, y)$ if j is 1 or 3, and as $m_{ij}(t, x)$ if j is 2 or 4. Describing the spatial

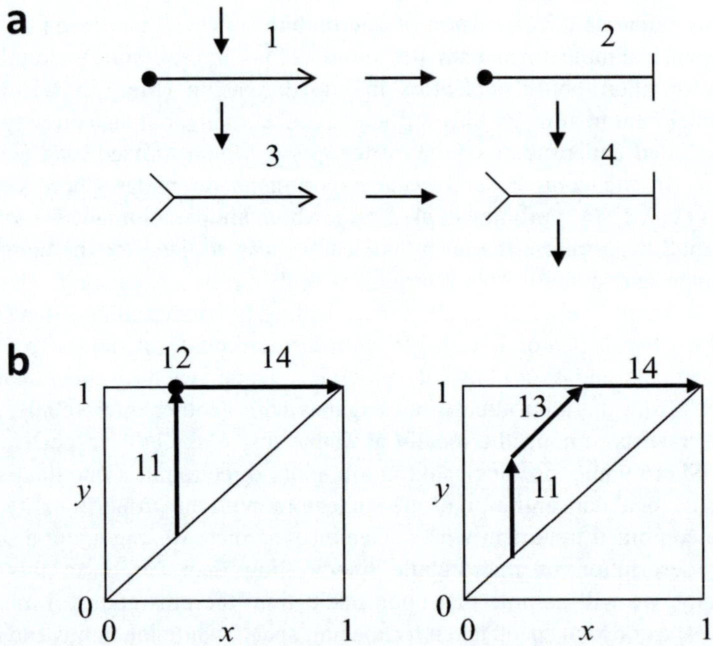

Fig. 5.8 (**a**) Dynamic states of a treadmilling microtubule and transitions between them. *Dot*, stable minus end, *barbed end*, disassembling minus end, *open arrowhead*, growing plus end, *vertical line*, boundary arresting the plus end, *arrows with filled arrowheads*, transitions between the numbered states. (**b**) Sample trajectories (*arrows*) of a microtubule in the space of the positions of its plus and minus ends. Successive dynamic types of the microtubule are indicated numerically. (Reproduced from Maly 2002 with permission)

redistribution of the nucleators, it will be necessary to distinguish between their motility on microtubules with stable and unstable minus ends. For this purpose, the state of the minus end of a microtubule along which a nucleator is moving will be given index 1 if this end is stable, and 2 if the end is disassembling. Further, the direction of movement of a nucleator that coincides with the direction of the coordinate axis will numbered 1, and the opposite direction of movement, 2. The state of the nucleator is indexed kl, where k is the direction of movement and l is the state of the minus end toward which the nucleator is moving. Then the distribution of the nucleators is described by number density functions $p_{kl}(t, x, z)$, where x is the coordinate of the minus end toward which the nucleator is moving and z is the coordinate of the nucleator.

The total concentration of the nucleators at a given location is

$$p_{\Sigma}(t,z) = \int_{1}^{z}\left[p_{11}(t,x,z) + p_{12}(t,x,z)\right]dx + \int_{z}^{0}\left[p_{21}(t,x,z) + p_{22}(t,x,z)\right]dx.$$

It determines nucleation of new microtubules symmetrically with respect to their orientation, because purely molecular-scale processes in the cytosol proceed isotropically:

$$m_{11}(t,z,z) = \frac{1}{2\sqrt{2}} p_{\Sigma}(t,z)n, \quad m_{21}(t,z,z) = \frac{1}{2\sqrt{2}} p_{\Sigma}(t,z)n.$$

Growth of the nucleated microtubules and destabilization of their minus ends determine the following dynamics of their distributions:

$$\frac{\partial m_{11}(t,x,y)}{\partial t} = -\frac{\partial m_{11}(t,x,y)}{\partial y} - rm_{11}(t,x,y), \quad x \le y,$$

$$\frac{\partial m_{21}(t,x,y)}{\partial t} = \frac{\partial m_{21}(t,x,y)}{\partial y} - rm_{21}(t,x,y), \quad x \ge y.$$

Arrival of the plus ends of state 1 microtubules at the boundary creates microtubules in dynamic state 2, which can similarly undergo minus end destabilization:

$$\frac{dm_{12}(t,x)}{dt} = m_{11}(t,x,1) - rm_{12}(t,x), \quad \frac{dm_{22}(t,x)}{dt} = m_{21}(t,x,0) - rm_{22}(t,x).$$

Dynamics in state 3 is given by

$$\frac{\partial m_{13}(t,x,y)}{\partial t} = -\frac{\partial m_{13}(t,x,y)}{\partial x} - \frac{\partial m_{13}(t,x,y)}{dy} + rm_{11}(t,x,y), \quad x < y,$$

$$\frac{\partial m_{23}(t,x,y)}{\partial t} = \frac{\partial m_{23}(t,x,y)}{\partial x} + \frac{\partial m_{23}(t,x,y)}{dy} + rm_{21}(t,x,y), \quad x > y,$$

$$m_{13}(t,0,y) = 0, \quad m_{23}(t,1,y) = 0,$$

and that in state 4, by

$$\frac{\partial m_{14}(t,x)}{\partial t} = -\frac{\partial m_{14}(t,x)}{\partial x} + m_{13}(t,x,1) + rm_{12}(t,x), \quad m_{14}(t,0) = 0,$$

$$\frac{\partial m_{24}(t,x)}{\partial t} = \frac{\partial m_{24}(t,x)}{\partial x} + m_{23}(t,x,0) + rm_{22}(t,x), \quad m_{24}(t,1) = 0.$$

Proceeding now to the dynamics of the nucleators, we will note, first, that dynein ferrying the nucleators to the minus ends is a processive molecular motor under intracellular conditions. Accordingly, the nucleator retains the velocity imparted to it by dynein up to the point at which it reaches the minus end of the microtubule that the dynein is attached to. The flux of nucleators off the minus ends at position z is

$$f(t,z) = v\left[p_{11}(t,z,z) + p_{21}(t,z,z)\right] + (v+1)\left[p_{12}(t,z,z) + p_{22}(t,z,z)\right],$$

where the second term accounts for the relative velocity of the nucleators and disassembling minus ends. This flux redistributes between all microtubules passing through the said point. The total number of microtubules passing through point z is

$$m_\Sigma(t,z) = \int_1^z \int_z^0 \left[m_{11}(t,x,y) + m_{13}(t,x,y)\right] dx dy + \int_z^0 \left[m_{12}(t,x) + m_{14}(t,x)\right] dx$$
$$+ \int_z^0 \int_1^z \left[m_{21}(t,x,y) + m_{23}(t,x,y)\right] dx dy + \int_1^z \left[m_{22}(t,x) + m_{24}(t,x)\right] dx.$$

Using this result, we can write

$$\frac{\partial p_{11}(t,x,z)}{\partial t} = f(t,z)\frac{\int_0^z m_{21}(t,x,y)dy + m_{22}(t,x)}{m_\Sigma(t,z)} - v\frac{\partial p_{11}(t,x,z)}{\partial z}$$
$$-rp_{11}(t,x,z), \quad x \geq z, \quad p_{11}(t,x,0) = 0,$$
$$\frac{\partial p_{12}(t,x,z)}{\partial t} = f(t,z)\frac{\int_0^z m_{23}(t,x,y)dy + m_{24}(t,x)}{m_\Sigma(t,z)} - v\frac{\partial p_{12}(t,x,z)}{\partial z}$$
$$+rp_{11}(t,x,z) + \frac{\partial p_{12}(t,x,z)}{\partial x}, \quad x \geq z, \quad p_{12}(t,x,0) = 0, \quad p_{12}(t,1,z) = 0$$

for the nucleators moving to the right and

$$\frac{\partial p_{21}(t,x,z)}{\partial t} = f(t,z)\frac{\int_z^1 m_{11}(t,x,y)\,dy + m_{12}(t,x)}{m_\Sigma(t,z)} + v\frac{\partial p_{21}(t,x,z)}{\partial z}$$

$$-rp_{21}(t,x,z), \quad x \le z, \quad p_{21}(t,x,1) = 0,$$

$$\frac{\partial p_{22}(t,x,z)}{\partial t} = f(t,z)\frac{\int_z^1 m_{13}(t,x,y)\,dy + m_{14}(t,x)}{m_\Sigma(t,z)} + v\frac{\partial p_{22}(t,x,z)}{\partial z}$$

$$+rp_{21}(t,x,z) - \frac{\partial p_{22}(t,x,z)}{\partial x}, \quad x \le z, \quad p_{22}(t,x,1) = 0, \quad p_{22}(t,0,z) = 0$$

for the nucleators moving to the left. In each of these partial differential equations, the first term accounts for the random engagement of nucleators with microtubules, the second for the nucleator transport, the third for destabilization of minus ends, and the fourth, present in the equations for nucleators moving on microtubules with destabilized minus ends, for minus end depolymerization. It may be unexpected to represent aspects of microtubule dynamics in equations for transport, but this makes possible a compact and transparent formalism for what might otherwise appear to permit only a simulational approach (transport along microtubules that are themselves dynamic).

The orientational alignment of microtubules can be measured by a polarity index defined as the difference between the numbers of microtubules oriented in each direction, expressed relative to the total number of microtubules that pass through the given position in the cytoplasmic domain:

$$q(t,z) = \{\int_1^z\int_z^0 [m_{11}(t,x,y) + m_{13}(t,x,y)]\,dxdy + \int_z^0 [m_{12}(t,x) + m_{14}(t,x)]\,dx$$

$$-\int_z^0\int_1^z [m_{21}(t,x,y) + m_{23}(t,x,y)]\,dxdy - \int_1^z [m_{22}(t,x) + m_{24}(t,x)]\,dx\} / m_\Sigma(t,z).$$

Such a polarity index is equal to one, where all microtubules point with their plus ends in the direction of the coordinate axis, to minus one, where all microtubules have the opposite orientation, and to zero, where microtubules pointing in the two directions are present in equal numbers.

There are two types of initial conditions that describe the start points of the experiments with microtubule array self-organization. First, there is the situation of random distribution of microtubules and nucleators in the domain: $m_{13}(0, x, y) = 1$, $x \le y$; $m_{23}(0, x, y) = 1$, $x \ge y$; $p_{12}(0, x, z) = 1$, $x \ge z$; $p_{22}(0, x, z) = 1$, $x \le z$; $m_{11}(0, x, y) = 0$, $m_{21}(0, x, y) = 0$, $m_{12}(0, x) = 0$, $m_{22}(0, x) = 0$, $m_{14}(0, x) = 0$, $m_{24}(0, x) = 0$, $p_{11}(0, x, z) = 0$, $p_{21}(0, x, z) = 0$. Another initial condition is that of the uniform orientation of microtubules stretching across the entire domain and nucleators concentrated at their minus ends: $m_{13}(0, x, y) = \delta(x)\delta(y{-}1)$, $p_{22}(0, x, z) = \delta(x)$, $m_{23}(0, x, y) = 0$, $p_{12}(0, x, z) = 0$, $m_{11}(0, x, y) = 0$, $m_{21}(0, x, y) = 0$, $m_{12}(0, x) = 0$, $m_{22}(0, x) = 0$, $m_{14}(0, x) = 0$, $m_{24}(0, x) = 0$, $p_{11}(0, x, z) = 0$, $p_{21}(0, x, z) = 0$.

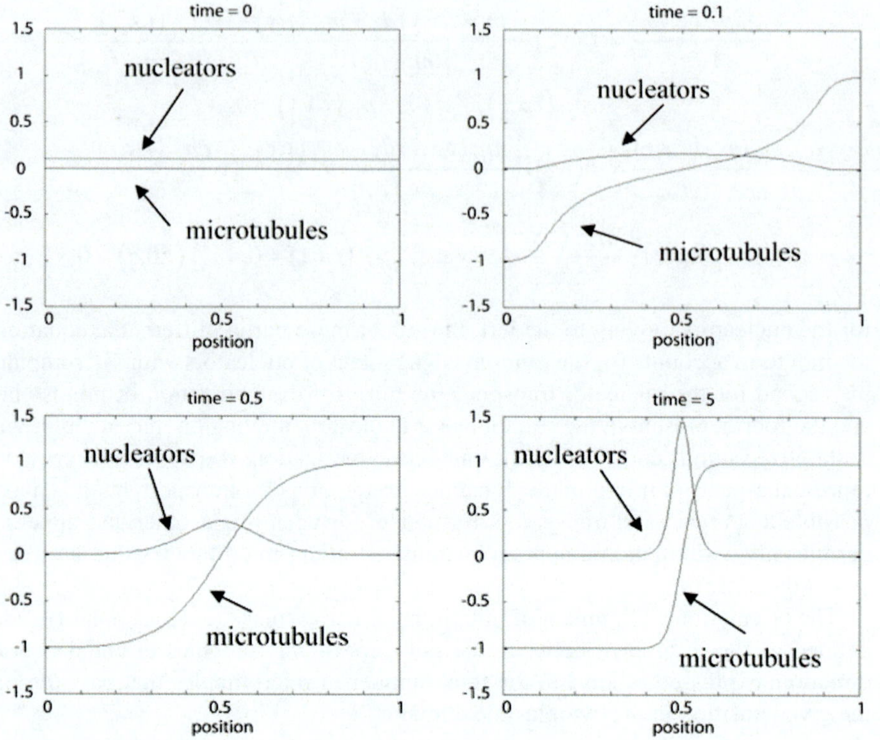

Fig. 5.9 Concentration of nucleators p_Σ and microtubule polarity index q at the specified times in a numerical solution of the microtubule array self-organization model. (Reproduced from Maly 2002 with permission)

Evolution of the system from the random initial condition is shown in Fig. 5.9. The plotted example is for $v = 10$, $r = 5$, and $n = 5$ (the units are set by the choices of the natural length and velocity that were made above—namely, the domain length and apparent velocity of dynamic microtubule ends). Other parameter values lead to the same qualitative behavior. The computation shows that the system spontaneously evolves from the random state to the state where nucleators are tightly concentrated in the middle of the cytoplasmic domain, and where microtubules in the right part of the domain point with their plus ends to the right, whereas microtubules in the left part point to the left. This arrangement is the same as the one that would result from the microtubules being organized by a centrosome positioned in the middle of a long, narrow domain of cytoplasm. Computations show that the system evolves to the same state from the other (unidirectional) initial condition as well. Thus, the combination of microtubule treadmilling with transport of microtubule nucleators to the minus ends is capable of producing a refined form of a centrally focused cytoskeleton.

5.3 Biochemistry of Cell Polarity

Although the most obvious mechanisms of self-organization leading to the establishment of the functional cell form arise from the motility and self-assembly of the cell's structural elements such as the cytoskeleton, spatial gradation of concentration of biochemical components also plays a role, as has been found in recent studies (e.g., Martin 2015; Prabhakar et al. 2020). Frequently, such gradation is achieved in cooperation with the directional cytoskeleton and motility components (Wedlich-Soldner et al. 2003; Ozbudak et al. 2005; Aguilar-Aragon et al. 2020) but, like the latter, the biochemical signaling machinery is capable of spatially self-organizing on its own. Even though the elements of the signaling machinery lack the ability to self-assemble into intrinsically directional supramolecular complexes which only need a higher-order spatial organization mechanism to acquire directionality on the cell scale, their local interactions combined with diffusion are capable of generating a cell-scale pattern. One instructive instance where the experiments suggest a central role for such a reaction-diffusion mechanism in higher eukaryotes involves autocrine signaling by EGF, a paradigmatic and multifunctional process that we have already encountered in the chapter on control. In this section, we will speak of cell polarization with the meaning of emergence in a cell of functionally distinct spatial regions, for example, front and back of a locomoting cell. Protrusive dynamics of the actin cortex being under control of EGF, the spontaneous polarization in the instance linked to EGF signaling manifests itself as acquisition of directionality in locomotion (Maheshwari et al. 2001). Specifically, autocrine secretion of EGF plays a critical role in the emergence of spatial localization of activity in the EGF signaling circuit (Joslin et al. 2007).

Spatial self-organization based on diffusion or diffusion-like mechanisms is known in various physicochemical and biological systems and is usually thought to stipulate two conditions: short-range activation and long-range inhibition (Turing 1952; Gierer and Meinhardt 1972). The principle of short-range activation and long-range inhibition has been applied to the emergence of cell polarity extensively—sometimes in abstract terms or implicitly (Meinhardt 1999; Sambeth and Baumgaertner 2001; Wedlich-Soldner et al. 2003), and at other times to guide model-building for a specific molecular network (Narang et al. 2001). In application to autocrine EGF signaling (Maly et al. 2004a), it has helped to identify the system's features that support self-organization of cell polarity.

In autocrine EGF signaling, ligand is shed from the cell surface and binds to transmembrane receptors that initiate an intracellular biochemical cascade that leads, ultimately, to stimulation of the ligand release. As already considered in the last chapter, recruitment of proteins such as Grb2 and SOS to the cytoplasmic domains of activated receptors forms the first step of the signal transduction in the cytoplasm, followed by activation of the MAPK cascade. Activated ERK at the end of the cascade is responsible for stimulating ligand release from the cell surface in cells that possess autocrine capabilities. Of note for the consideration of the spatial range of the autocrine circuit, the receptor complexes display only a limited mobility in the membrane. As a result, the signal is propagated in space primarily by the ligand outside the cell and by active phosphorylated forms of the MAP kinases in the cytoplasm. Spatial propagation of signal by EGF family ligands and MAP kinases is limited by the kinetics of their re-capture and dephosporylation, respectively (Lauffenburger et al. 1998; Brown and Kholodenko 1999). Hence, activation processes in the autocrine circuit can be localized, supplying one of the requirements for spatial self-organization.

How is the other requirement, global inhibition, realized in this molecular system? The forms of inhibition of EGFR signaling that are based on inactivation of receptor complexes, such as dephosphorylation of cytoplasmic domains of activated receptors or endocytosis of the receptor complexes, affect only signaling complexes that are already active. Phosphorylation of SOS, which is promoted by activated ERK (see last chapter), is another inhibitory element in the autocrine network. The phosphorylated form of SOS is a weak participant of the transmembrane signaling complexes, and also a freely diffusible cytoplasmic protein, like the unphosphorylated form. All that is required for SOS phosphorylation in the cell region where signaling is strong to consume free unphosphorylated SOS in a larger region is that the conversion between the two forms be slow relative to diffusion. The lowered availability of SOS for the signaling complex formation can inhibit signal transduction, potentially, throughout the cell. The ERK-SOS negative feedback therefore is capable of supplying global inhibition as the second condition for spatial self-organization.

Another mechanism of inhibition that limits signaling is depletion of inactive forms of signaling proteins. A cytoplasmic component of the transmembrane signaling complex, for example, Grb2 or SOS, changes its mobility significantly upon becoming part of the large, membrane-bound complex. Thus, while it contributes to the signal propagation only near the point of its activation, its incorporation into the active signaling complex can result in lowering the concentration of its free form everywhere in the cell. Depletion of cytoplasmic components of transmembrane signaling complexes is a second likely mechanism of global spatial inhibition in the EGF autocrine network.

To analyze the origin of spatial self-organization in this system, the dynamics of autocrine EGF signaling (Shvartsman et al. 2002) have been integrated with the kinetics of the Shc-Grb-SOS complex formation and MAPK phosphorylation cascade in a partial differential equations formalism (Maly et al. 2004a). The intrinsic, non-spatial dynamics the MAPK cascade with negative feedback to SOS have been treated in relation to oscillatory and switch-like control in the previous chapter. The autocrine system model additionally accounts for the MAPK-activated shedding of EGF from the surface of the cell, which constitutes positive feedback to the very root of the EGFR signaling cascade (Fig. 5.10). The essential steps that have to be accounted for explicitly are ligand shedding, receptor binding, assembly of the receptor signaling complex by recruitment of cytoplasmic Grb2 and SOS, phosphorylation of Raf, and two-site phosphorylation of both MEK and ERK MAP kinases. It can be assumed, as a simplification, that the ERK phosphorylation of

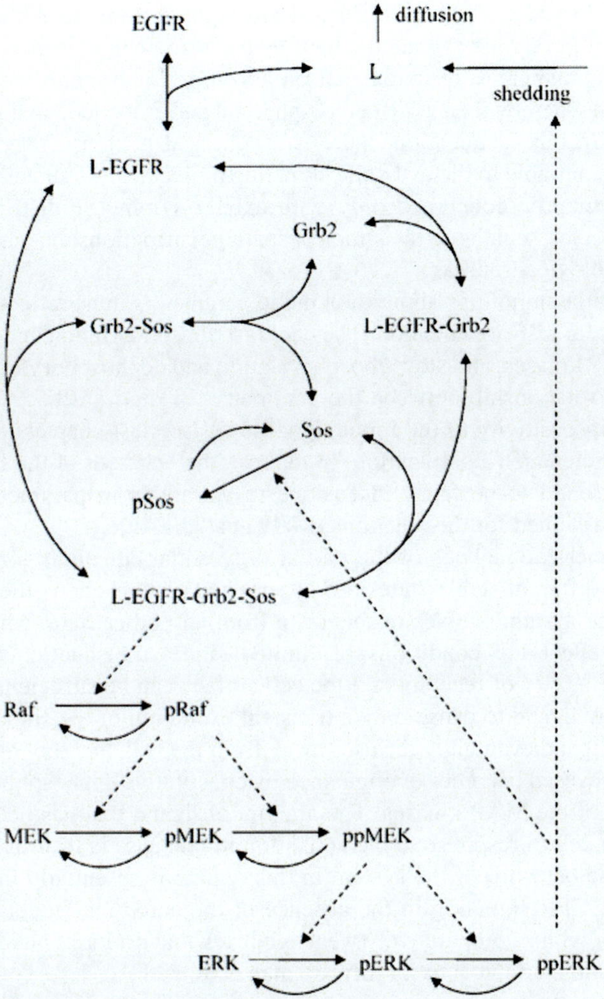

Fig. 5.10 Autocrine EGFR signaling. *L*, ligand, *p*, phosphorylated form, *pp*, doubly phosphorylated form. *Solid arrows*: transport, covalent modifications, and protein binding processes. *Dashed arrows*: kinetic acceleration of a process by a chemical species. (Reproduced from Maly et al. 2004a with permission)

SOS renders the latter completely incapable to participate in the receptor signaling complex formation.

The molecular processes depicted in Fig. 5.10 occur in three spatial dimensions, two separate compartments, and on the boundary between the latter. EGFR ligand diffuses in the medium beyond the cell membrane, and the intracellular components, namely Grb2, Grb2-SOS, and the different forms of SOS, Raf, MEK, and ERK, diffuse in the cytoplasm. The receptor and its complexes with EGF-family ligands, Grb2, and SOS can be treated as fixed in the plasma membrane, because of the relatively low rates of lateral diffusion within the membrane. Although behavior of this system in complex tissue geometries has also been considered (Maly et al. 2004b), the fundamentals of spatial self-organization can be analyzed by adopting a simplification whereby the single cell is spherical and embedded in an infinite external volume. In this idealization, there is a spherical nucleus in the cell's center, which is impenetrable to the cytosolic signaling molecules. As an additional simplification, we may be concerned only with axially symmetric distributions of the molecular species, which are the simplest spatial distributions that can be polarized in the cell-biological sense.

The crucial assumptions about the kinetic parameters must reflect the discussed requirements for self-organization. They are fast phosphorylation and dephosphorylation of MAP kinases, and slow phosphorylation and dephosphorylation of SOS. In addition, the relationship between the parameters of the MAPK cascade that provides for ultrasensitivity to the input signal level (see last chapter) must be maintained for efficient self-organization. In the rest, the behavior of the EGF autocrine system was found to be fairly insensitive to variations in parameter values and kinetic laws assumed for the reactions (Maly et al. 2004a).

In the numerical analysis of the partial differential equations formalism, three qualitatively different stable states are observed. The first one is the state with no signaling. The system evolves to this state from any other state, when the kinetic parameters reflect the conditions of limited signal transduction potential. For instance, the density of receptors on the cell surface can be sufficiently low for the stimulation by ligand to cause only a transient excitation of the signaling network. The applied ligand eventually diffuses away from the cell, and the intracellular signaling that it caused becomes extinguished by constitutive dephosphorylation of the active forms of the MAP kinases. The amount of ligand that has been released by the autocrine cell in response to the activation in this case is insufficient to sustain signaling. The behavior of the system in this regime is essentially the same as the behavior of EGFR signaling in the absence of the autocrine feedback (Schoeberl et al. 2002) or with a comparatively weak feedback that prolongs but does not indefinitely sustain the circuit activity (Shvartsman et al. 2002).

The system adopts a qualitatively different kind of stable state when the parameter values signify a higher signaling potential, for instance, when the surface density of receptors is high. This is the stable state wherein autocrine signaling is sustained and distributed uniformly over the cell surface. Ligand is shed from the surface everywhere at an equal rate, causing an equal stimulation of all the plasma membrane receptors. Diffusion of ligand and of the activated forms of intracellular proteins from the cell membrane under these conditions results in spherically symmetric spatial distributions (Fig. 5.11a). The sustained and spherically symmetric signaling is of the same type as the spherically symmetric extracellular ligand distribution modeled by Shvartsman et al. (2002).

In addition, the system demonstrates a third kind of stable state, whereby the ligand is released predominantly from one side of the cell, and the receptor activation and intracellular signaling are concentrated on that "active" side of the cell. Near the opposite, "inactive" pole of the cell, ligand shedding, receptor activation,

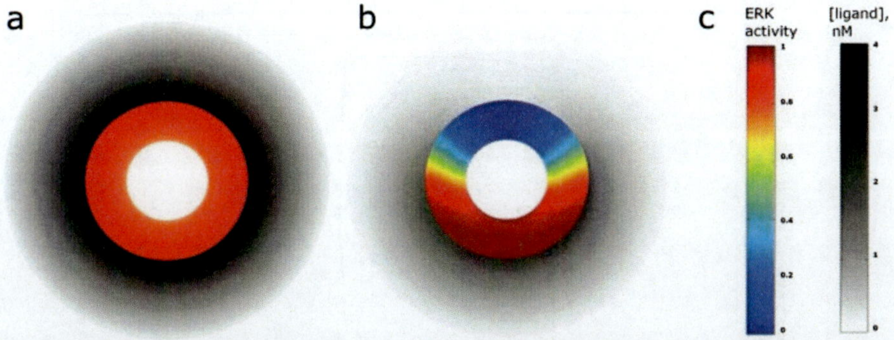

Fig. 5.11 Axial sections through stable spatial distributions in the autocrine EGF signaling system. (**a**) Spherically symmetric case, (**b**) Self-organized case, (**c**) Concentration scales. The ERK activity is measured by the fraction of its doubly phosphorylated form. The boundary between the ERK and ligand shaded zones is the cell surface; the non-shaded center is the nucleus. Reproduced from Maly et al. (2004a). Reprinted from Biophysical Journal, Vol. 86, Ivan V. Maly, H. Steven Wiley, Douglas A. Lauffenburger, Self-Organization of Polarized Cell Signaling via Autocrine Circuits: Computational Model Analysis, 10–22, copyright 2004, with permission from Elsevier

and intracellular signaling are maintained at low levels (Fig. 5.11b). This type of spatial distribution possesses only axial symmetry, with the symmetry axis passing through the two emergent poles, where the levels of signaling are maximal and minimal. Such a polarization of an otherwise spherically symmetric cell reflects self-organization in the spatially distributed signaling network. It is achieved when the kinetic parameters signify an intermediate potential for signal transduction. For instance, the system self-organizes in space when the surface density of receptors is between the high and low values that lead respectively to the uniformly high or uniformly zero asymptotic levels of autocrine signaling.

The autocrine system undergoes bifurcations causing abrupt transitions between the different types of stable signaling distributions, as parameters reach certain threshold values. For instance, the kinetic conditions may be such that autocrine signaling is not sustained if the density of receptors at the cell surface falls below ~400 per square micron (Fig. 5.12). Quasi-static increase of the receptor density above this value causes a rapid transition to the self-organized state of spatially polarized signaling. As shown in Fig. 5.12, this state is characterized by divergent levels of EGF receptor occupancy at the opposite poles of the cell. At one pole (which emerges on the idealized spherical cell solely due to self-organization of signaling), most receptors are occupied by ligand and transmit the signal to the underlying cytoplasm. Through the positive feedback loop in the signaling network, this leads to secretion of more ligand at that pole, and the local autocrine signaling is maintained indefinitely at a high level. At the opposite pole, only a small fraction of receptors is occupied by ligand (Fig. 5.12), which results in a locally low level of autocrine signaling. Further increase in the total receptor density does not result in any dramatic changes of the spatial distribution of receptor occupancy or signaling until another threshold is reached at ~800 receptors per square micron of the plasma

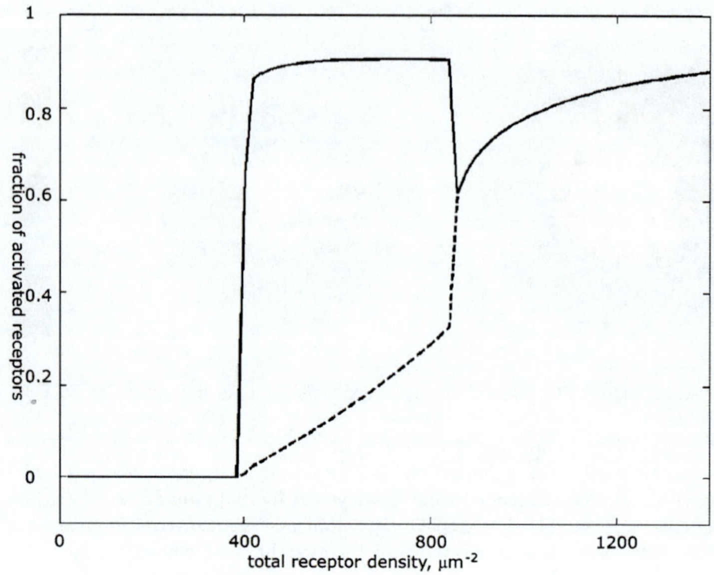

Fig. 5.12 Receptor occupancy during quasi-static variation of the total receptor density. *Solid* and *dashed curves* trace the occupancy at the opposite sides of the cell. Reproduced from Maly et al. (2004a). Reprinted from Biophysical Journal, Vol. 86, Ivan V. Maly, H. Steven Wiley, Douglas A. Lauffenburger, Self-Organization of Polarized Cell Signaling via Autocrine Circuits: Computational Model Analysis, 10–22, copyright 2004, with permission from Elsevier

membrane. Quasi-static increase of the density beyond this value causes the self-organized state to collapse, the receptor occupancy dropping at the active pole and rising at the inactive pole to equalize at a relatively high level. Above this second threshold, the steady-state occupancy of receptors becomes, once again, relatively independent of the total receptor density. The system displays a similar three-stage, two-threshold response to quasi-static variation of its other parameters.

Determination of the threshold (bifurcation) surfaces that divide the multidimensional parameter space reveals some interesting relationships between the cellular parameters. One such relationship involves the receptor density and the rate constant of ligand shedding, which control the spatial range of autocrine activation. Higher rate of ligand release increases the ligand concentration not only locally, but also, due to diffusion, at the plasma membrane regions that are removed from the point where the ligand is released. The effect of diffusion ensures that a lower rate of release will increase the ligand concentration mostly locally, as the steady-state profile of concentration drops sharply with the distance from the source in three dimensions (Shvartsman et al. 2001). A higher density of receptors at the membrane, however, enables a stronger stimulation of intracellular signaling even by the lower concentration of ligand, extending the spatial range of the autocrine activation. A section through the parameter space reveals that when both the receptor density and the rate of shedding are low, signaling cannot be sustained, and when

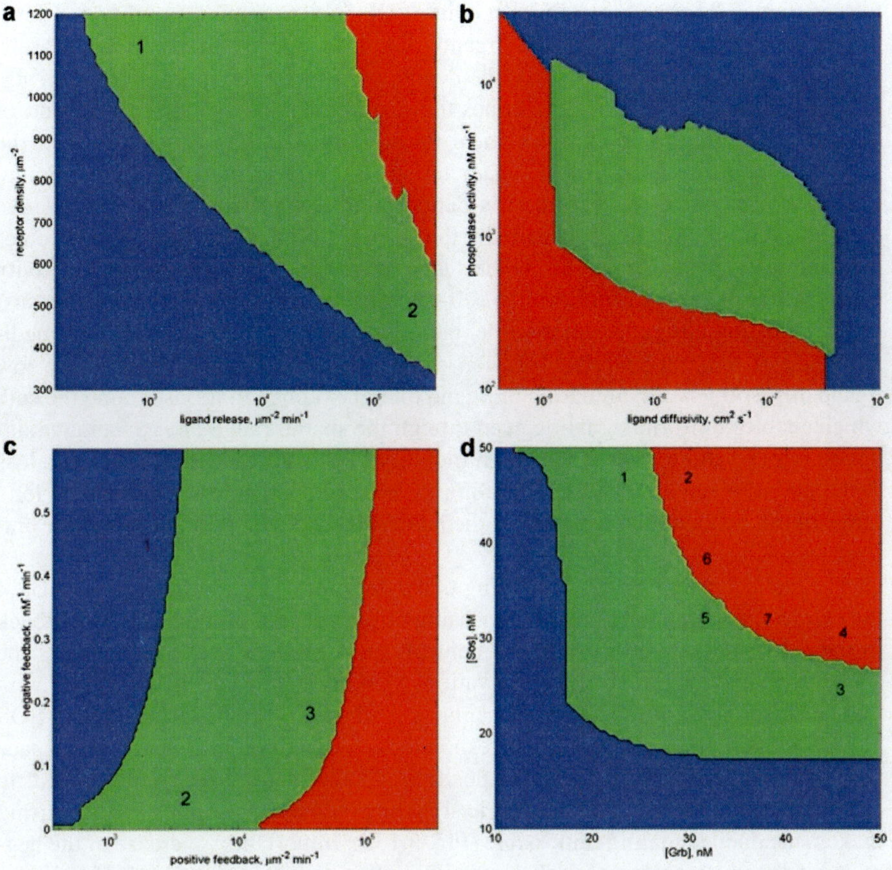

Fig. 5.13 Parameter space of autocrine EGF signaling. (**a**) Receptor density *vs.* maximum rate of ligand release, (**b**) Phosphatase activity *vs.* ligand diffusivity, (**c**) Negative *vs.* positive feedback strength (maximum rate of ligand release *vs.* SOS phosphorylation rate constant), (**d**) Expression of SOS *vs.* Grb2 (measured as formal concentrations in the total cell volume). *Blue*, no signaling, *red*, uniform signaling, *green*, spatially self-organized signaling at steady state. Numbers mark parameter space regions discussed in the text. Reproduced from Maly et al. (2004a). Reprinted from Biophysical Journal, Vol. 86, Ivan V. Maly, H. Steven Wiley, Douglas A. Lauffenburger, Self-Organization of Polarized Cell Signaling via Autocrine Circuits: Computational Model Analysis, 10–22, copyright 2004, with permission from Elsevier

they are high, activation spreads over the entire cell (Fig. 5.13a). The spatially self-organized state of signaling requires certain intermediate values of these parameters. A low rate of shedding accommodates relatively high receptor densities (region 1 in Fig. 5.13a), whereas a high shedding rate must be accompanied by a low receptor density for the self-organization to take place (region 2 in Fig. 5.13a).

The spatial range of the extracellular part of the autocrine circuit is also controlled by the diffusion constant of the secreted ligand. The diffusion constant of growth factors can vary over several orders of magnitude, depending on the microstructure of the extracellular medium. High extracellular mobility can level the concentration profile of the ligand around the cell, making spatial self-organization impossible. Similarly, the degree of intracellular mobility of such signaling molecules as MAPK determines the spatial range of the intracellular part of the autocrine circuit and can be critical for self-organization. More precisely, the spatial scale is determined by the characteristic distance an activated MAPK molecule covers by diffusion before it is dephosphorylated (Brown and Kholodenko 1999), which is controlled primarily by the activity of the phosphatase. A section of the parameter space (Fig. 5.13b) reveals that an increase in the phosphatase activity causes a transition from uniform to self-organized signaling before it makes sustained signaling altogether impossible, but only does so within a certain intermediate range of the extracellular diffusion constant. The dominant effect of the low ligand diffusivity is accumulation of ligand to higher concentrations around the cell, which enables uniform signaling even though the spatial range of extracellular signaling may be short. Conversely, high mobility of ligand leads primarily to its loss to the environment, and causes signaling to cease altogether, even though the long range of the extracellular segment of the autocrine loop can, in principle, favor uniform signaling.

The rate constant of ligand shedding that determines, in part, the spatial range of the extracellular signaling, is also a measure of the strength of the positive feedback in the autocrine network, because it controls the steepness of the response of the shedding rate to the level of ERK activation. As such, its role can be juxtaposed with that of the strength of the negative feedback mediated by the ERK-dependent SOS phosphorylation (Fig. 5.10). The correspondent section through the parameter space shows that a combination of a strong positive and a weak negative feedback leads to establishment of spherically symmetric signaling, whereas the opposite relationship makes sustained signaling impossible (Fig. 5.13c). Importantly, increase of the negative feedback strength can abolish signaling altogether (as in region 1, Fig. 5.13c) or cause self-organization of signaling (as in region 3 of the same diagram), depending on the strength of the positive feedback. Although the negative feedback can cause self-organization, it is not required for it, as evidenced by the self-organization occurring in region 2 of the parameter space section in Fig. 5.13c, where the negative feedback is weak or absent.

The self-organization in the absence of the negative feedback through the ERK-dependent phosphorylation of SOS can be caused by limited overall availability of SOS or the other cytoplasmic component of the active receptor complex, Grb2. The respective section through the parameter space shows that high levels of expression of both SOS and Grb2 allow for a uniformly high level of autocrine signaling (Fig. 5.13d). A low level of expression of either of these proteins precludes sustained signaling, but at certain intermediate, limited levels of their expression, a spatially self-organized state is achieved. The availability of either protein can limit signaling to one side of the cell. The diagram in Fig. 5.13d demonstrates that at relatively high levels of expression of both proteins simultaneously, their complex that is required for signal transduction can still be limiting, enabling spatial self-organization in the system. Note that, for example, in the parameter space region marked “1” in Fig. 5.13d, Grb2 is limiting whereas the amount of SOS allows for uniform signaling in the nearby region marked “2.” Conversely, in region 3, only SOS is limiting, whereas the amount of Grb2 allows for uniform signaling in region 4. In region 5, it is the combination of the two proteins that is limiting, whereas the level of each protein individually is sufficient for uniform signaling in neighboring regions 6 and 7.

On the whole, examination of the parameter space reveals an extensive multidimensional domain of cellular conditions that support spatial self-organization of autocrine signaling. As noted by Shvartsman et al. (2001), spatial localization of steady-state signaling requires that the activation loop of the molecular interaction network be localized. In the system displaying spontaneous self-organization, as laid out above, this condition is satisfied primarily due to the limited rate of autocrine ligand release and limited receptor density. The complementary conditions of global inhibition are satisfied by a relatively low level of expression of the Ras nucleotide exchange factor SOS, relatively strong negative feedback in the signaling system, namely the ERK-dependent SOS phosphorylation, and comparatively low expression of the SH2-SH3 domain adaptor protein Grb2 that recruits SOS to receptor complexes at the plasma membrane. The numerical analysis (Maly et al. 2004a) shows that every one of these conditions of global inhibition can be sufficient, and also that these factors may act synergistically and lead to self-organization when any one of them is quantitatively insufficient by itself. The fact that no other global inhibition mechanisms relevant to self-organization are constituted in the considerably complex biomolecular interaction network (Fig. 5.10) is remarkable, even though, of course, our improving understanding of autocrine signaling kinetics is likely to add to this picture.

The three global inhibition factors pinpointed here limit the signaling capacity of the cell, and as such can lead to the cell’s inability to sustain autocrine signaling if they become too strong. The numerical analysis shows, however, that as the spatially distributed system becomes inhibited by these factors, the signaling does not diminish monotonically and uniformly throughout the cell. Rather, as the inhibition exceeds a certain threshold, signaling at one side of the cell drops, whereas at the opposite pole it increases. Thus, the system differentiates spatially when its self-sustained activity cannot be maintained uniformly high due to the specific kinetic limitations on signal transduction.

It should be noted that the axial symmetry of the considered self-organized distributions results from instability of the spherically symmetrical ones, but more complex stable distributions are possible. Moreover, the sources of the defects of the spherical symmetry that are seeds of the axially-symmetric polarized state of the system can be diverse and include shallow but systematic gradients in extracellular signaling influences and anisotropies in the internal cellular organization. These sources may also derive from the histological organization of the tissue in the organism. Thus, capacity for spontaneous self-organization in the cell signaling, while it can manifest itself in isolated cells functioning in the homogeneous environment of the laboratory experiment, should function more frequently as a means to detect and amplify weak external signals and align the cell’s activity with the chemical and physical gradients in a real environment. It can also spatially align the cell's development and interaction with its immediate neighbors with morphogenetic gradients and tissue anisotropy in a multicellular organism. At present, there remains a considerable gap in understanding of such molecular mechanisms, which tie the cell-level spatial organization mechanisms with the organism-level morphogenetic mechanisms that will be considered in a subsequent chapter.

References

Abramowitz M, Stegun IA (eds) (1972) Handbook of mathematical functions with formulas, graphs, and mathematical tables. Dover, New York

Aguilar-Aragon M, Fletcher G, Thompson BJ (2020) The cytoskeletal motor proteins dynein and MyoV direct apical transport of crumbs. Dev Biol 459:126–137

Boveri T (1887) Ueber den Antheil des Spermatozoon an der Teilung des Eies. Sitzungsber Ges Morph Physiol München 3:151–164

Bray D (2001) Cell movements: from molecules to motility. Garland, New York

Brown GC, Kholodenko BN (1999) Spatial gradients of cellular phospho-proteins. FEBS Lett 457:452–454

Cytrynbaum EN, Rodionov V, Mogilner A (2004) Computational model of dynein-dependent self-organization of microtubule asters. J Cell Sci 117:1381–1397

Cytrynbaum EN, Rodionov V, Mogilner A (2006) Nonlocal mechanism of self-organization and centering of microtubule asters. Bull Math Biol 68:1053–1072

Eigen M, McCaskill J, Schuster P (1989) The molecular quasispecies. Adv Chem Phys 75:149–263

Fisher RA (1999) The genetical theory of natural selection. Oxford University Press, Oxford

Gierer A, Meinhardt H (1972) A theory of biological pattern formation. Kybernetik 12:30–39

Joslin EJ, Opresko LK, Wells A et al (2007) EGF-receptor-mediated mammary epithelial cell migration is driven by sustained ERK signaling from autocrine stimulation. J Cell Sci 120:3688–3699

Kirschner M, Gerhart J, Mitchison T (2000) Molecular "vitalism.". Cell 100:79–98

Lauffenburger DA, Oehrtman GT, Walker L et al (1998) Real-time quantitative measurement of autocrine ligand binding indicates that autocrine loops are spatially localized. Proc Natl Acad Sci U S A 95:15368–15373

Maheshwari G, Wiley HS, Lauffenburger DA (2001) Autocrine epidermal growth factor signaling stimulates directionally persistent mammary epithelial cell migration. J Cell Biol 155:1123–1128

Malikov V, Cytrynbaum EN, Kashina A et al (2005) Centering of a radial microtubule array by translocation along microtubules spontaneously nucleated in the cytoplasm. Nat Cell Biol 7:1213–1218

Maly IV (2002) Spatial organization in kinetic models of the cytoskeleton and cell motility. Dissertation, Northwestern University

Maly IV, Borisy GG (2001) Self-organization of a propulsive actin network as an evolutionary process. Proc Natl Acad Sci U S A 98:11324–11329

Maly IV, Borisy GG (2002) Self-organization of treadmilling microtubules into a polar array. Trends Cell Biol 12:462–465

Maly IV, Wiley HS, Lauffenburger DA (2004a) Self-organization of polarized cell signaling via autocrine circuits: computational model analysis. Biophys J 86:10–22

Maly IV, Lee RT, Lauffenburger DA (2004b) A model for mechanotransduction in cardiac muscle: effects of extracellular matrix deformation on autocrine signaling. Ann Biomed Eng 32:1319–1335

Martin SG (2015) Spontaneous cell polarization: feedback control of Cdc42 GTPase breaks cellular symmetry. BioEssays 37:1193–1201

McNiven M, Porter KR (1986) Microtubule polarity confers direction of pigment transport in chromatophores. J Cell Biol 103:1547–1555

McNiven M, Porter KR (1988) Organization of microtubules in centrosome-free cytoplasm. J Cell Biol 106:1593–1605

McNiven M, Wang M, Porter KR (1984) Microtubule polarity and the direction of pigment transport reverse simultaneously in surgically severed melanophore arms. Cell 37:753–765

Meinhardt H (1999) Orientation of chemotactic cells and growth cones: models and mechanisms. J Cell Sci 112:2867–2874

Michod RE (1999) Darwinian dynamics: evolutionary transitions in fitness and individuality. Princeton University Press, Princeton, NJ

Mogilner A, Oster G (1996) Cell motility driven by actin polymerization. Biophys J 71:3030–3045

Narang A, Subramanian KK, Lauffenburger DA (2001) A mathematical model for chemoattractant gradient sensing based on receptor-regulated membrane phospholipid signaling dynamics. Ann Biomed Eng 29:677–691

Nedelec FJ, Surrey T, Maggs AC et al (1997) Self-organization of microtubules and motors. Nature 389:305–308

Nowak MA (1992) What is a quasispecies? Trends Ecol Evol 7:118–121

Nowak MA, Plotkin JB, Jansen VAA (2000) The evolution of syntactic communication. Nature 404:495–498

Nowak MA, Komarova NL, Niyogi P (2001) Evolution of universal grammar. Science 291:114–118

Ozbudak EM, Becskei A, van Oudenaarden A (2005) A system of counteracting feedback loops regulates Cdc42p activity during spontaneous cell polarization. Dev Cell 9:565–571

Page KM, Nowak MA (2002) Unifying evolutionary dynamics. J Theor Biol 219:93–98

Prabhakar A, Chow J, Siegel AJ et al (2020) Regulation of intrinsic polarity establishment by a differentiation-type MAPK pathway in *S. cerevisiae*. J Cell Sci 133:jcs241513

Rodionov VI, Borisy GG (1997a) Self-centring activity of the cytoplasm. Nature 386:170–173

Rodionov VI, Borisy GG (1997b) Microtubule treadmilling in vivo. Science 275:215–218

Rodionov VI, Nadezhdina E, Borisy GG (1999) Centrosomal control of microtubule dynamics. Proc Natl Acad Sci U S A 96:115–120

Rowe GW (1994) Theoretical models in biology: the origin of life, the immune system, and the brain. Oxford University Press, New York

Sambeth R, Baumgaertner A (2001) Autocatalytic polymerization generates persistent random walk of crawling cells. Phys Rev Lett 86:5196–5199

Schoeberl B, Eichler-Jonsson C, Gilles ED et al (2002) Computational modeling of the dynamics of the MAP kinase cascade activated by surface and internalized EGF receptors. Nat Biotechnol 20:370–375

Sharp DJ, Rogers GC, Scholey JM (2000) Roles of motor proteins in building microtubule-based structures: a basic principle of cellular design. Biochim Biophys Acta 1496:128–141

Shvartsman SY, Wiley HS, Deen WM et al (2001) Spatial range of autocrine signaling: modeling and computational analysis. Biophys J 81:1854–1867

Shvartsman SY, Hagan MP, Yacoub A et al (2002) Autocrine loops with positive feedback enable context-depending cell signaling. Am J Physiol Cell Physiol 282:C545–C559

Svitkina TM, Borisy GG (1999) Arp2/3 complex and actin depolymerizing factor/cofilin in dendritic organization and treadmilling of actin filament array in lamellipodia. J Cell Biol 145:1009–1026

Turing A (1952) The chemical basis of morphogenesis. Phil Trans Roy Soc B 237:37–72

Verde F, Berrez JM, Antony C et al (1991) Taxol-induced microtubule asters in mitotic extracts of *Xenopus* eggs: requirement for phosphorylated factors and cytoplasmic dynein. J Cell Biol 112:1177–1187

Vorobjev IA, Svitkina TM, Borisy GG (1997) Cytoplasmic assembly of microtubules in cultured cells. J Cell Sci 110:2635–2645

Vorobjev IA, Rodionov VI, Maly IV et al (1999) Contribution of plus and minus end pathways to microtubule turnover. J Cell Sci 112:2277–2289
Vorobjev IA, Malikov V, Rodionov VI (2001) Self-organization of a radial microtubule array by dynein-dependent nucleation of microtubules. Proc Natl Acad Sci U S A 98:10160–10165
Wedlich-Soldner R, Altschuler S, Wu L et al (2003) Spontaneous cell polarization through actomyosin-based delivery of the Cdc42 GTPase. Science 299:1231–1235

Chapter 6
Forces that Shape the Cell

6.1 Symmetry-Breaking and Emergent Irreversibility

Working on the already fairly intricately organized material supplied by intracellular self-organization, mechanical forces are responsible for generation of more complex form and movement that are the distinguishing characteristics of living matter. While the process of mechanically shaping matter in unmistakably biological ways extends to the organ and organism levels—to be addressed later in the book—its action is crucial also on the level of isolated or directly interacting cells. This hierarchy places systems biomechanics of the cell (Maly 2013) at the root of morphogenesis and motility that are not meaningfully reducible to even the complex self-organization effects in the molecular realm, but arise entirely as properties of the structure (cell boundary confining the cytoplasmic material inside) which, in developed life, is only inherited from cell to cell (Harold 2001).

In the present chapter, we will sketch out the most elementary effects in systems biomechanics of the cell, along with the approaches to their analysis. The microtubule cytoskeleton confined in the cell boundary supplies a sufficiently investigated example of this type of dynamics. Self-organization of its focal intrinsic configuration, whereby microtubules become connected in the form of a radiating aster, has been discussed in the last chapter, along with the functional physiological significance of the order and anisotropy that the aster arrangement confers to the rest of the eukaryotic cell's cytoplasm. Forces resulting from the aster's confinement within cell boundary, as we shall see, are capable of inducing asymmetry in the microtubule cytoskeleton. A special but biologically essential case of such asymmetry induction involves the mitotic microtubule cytoskeleton, which consists, in effect, of two mechanically coupled asters in preparation for cell division. In this instance, the cytoskeleton asymmetry becomes reflected in asymmetric cell division, which

This chapter is based closely on the author's prior work, Systems Biomechanics of the Cell (Springer, 2013).

I. Maly, *Quantitative Elements of General Biology*,
https://doi.org/10.1007/978-3-030-79146-9_6

underlies much of the morphogenetic differentiation in developing organisms. Finally, we will consider how the direction of the aster asymmetry becomes selected relative to cell-cell contacts. Although each of these processes may appear mechanically simple in the abstract, analysis shows that the physics of the confined microtubule cytoskeleton exhibits a series of system-level effects, which include—in addition to spontaneous symmetry-breaking—irreversibility, structural multistability, and mutiperiodic movement. Unidirectional development leading to complex form and movement being the hallmark of life, we will see that the effects arising in the mechanics of the confined cytoskeleton both exemplify it on the cellular level and supply the foundation for the hierarchy of dynamically similar effects in multicellular systems.

Proceeding to the most fundamental question of symmetry and asymmetry generation by cellular forces, let us recall from the last chapter that in the typical metazoan cell, the prominent structural feature is the aster of microtubules anchored on the centrosome (Bray 2001). The microtubules direct transport of organelles and secretory vesicles, making the position of the centrosome functionally important for various cellular activities, such as wound closure, migration, and interactions between cells of the immune system (Gotlieb et al. 1981; Kupfer et al. 1982; Kupfer and Singer 1989; Kupfer et al. 1994; Ueda et al. 1997; Abrahamsen et al. 2018). In particular, induction of asymmetry of the microtubule aster in some situations, and maintenance of its symmetry within the cell outline in others, has been a subject of intense experimental investigation (Kupfer et al. 1982; Euteneuer and Schliwa 1992; Koonce et al. 1999; Piel et al. 2000; Etienne-Manneville and Hall 2001; Yvon et al. 2002; Burakov et al. 2003; Gomes et al. 2005; Dupin et al. 2009; Tang and Marshall 2012; Koonce and Tikhonenko 2018).

Compact immune cells, such as lymphocytes—which have nearly spherical cell boundary outlines at least in some well-studied conditions, *e.g.*, in lymph or the typical experiment—present one of the simplest model cases for the mechanics of centrosome positioning. The eccentric centrosome position in lymphocytes is constitutive, and only its orientation to a specific side of the cell is regulated by the antigen-mediated cell-cell interaction. By contrast, the thinly spread epithelioid cells, which are the best studied experimentally, are nearly flat, with the exception of the nucleus, and supply a geometrically different model case.

Among the mechanisms implicated in centrosome positioning, we find microtubule polymerization dynamics (Stowers et al. 1995; Lowin-Kropf et al. 1998; Faivre-Moskalenko and Dogterom 2002; Yvon et al. 2002; Burakov et al. 2003), action of cortically-anchored molecular motors of the dynein type (Etienne-Manneville and Hall 2001; Burakov et al. 2003; Levy and Holzbaur 2008), movement of the entire cell that entrains the centrosome (Arkhipov and Maly 2006a, 2008), flow of cortical actomyosin that entrains microtubules (Burakov et al. 2003; Gomes et al. 2005), and even cell population-level kinetic selection linked to the direction of transport along the microtubules (Arkhipov and Maly 2006b, 2007; Kim 2019). Furthermore, a great number of biomolecules regulate the centrosome positioning, in the better characterized cases doing so by influencing one of the known positioning mechanisms. How general each mechanism and regulatory influence is between cell types, species, and developmental stages, and how exactly they interact or interfere with each other, is not a settled question and will, undoubtedly, remain a fruitful direction for future research. Integrative investigation will need to be quantitative and build sequentially on a solid conceptual foundation that takes into account the most general effects first. In addition to supplying the basis for understanding the variation of the cytoskeletal hub positioning in different organisms and cell types, such fundamental effects are of interest from the evolutionary and astrobiological viewpoint, because they can be presumed to be both primordial and generalizable beyond the realm presently open to cell-biological observation.

The most elementary contribution to centrosome positioning arises from bending elasticity of microtubules and represents an effect that is inseparable from the microtubule cytoskeleton itself. The fundamental role of the elastic compactization of the microtubule cytoskeleton within the constraints of the cell boundary for centrosome positioning was recognized early by Holy. By considering the absolute energy minimum of an aster consisting of a large finite number of evenly spaced microtubules of equal length, which was confined to a flat circular domain, it was shown that when the microtubule length exceeds the confining radius, the centrosome would become positioned eccentrically (Holy 1997). The equilibrium centrosome position in this idealization was a function of the length and radius, and a quantitative argument was put forward that the instability of the central position should be stronger in three than in two dimensions.

Subsequent development of the centrosome positioning theory incorporated, firstly, stochastic changes in the lengths of individual microtubules (Holy et al. 1997), which occur, as was briefly reviewed in the last chapter, in the course of tubulin polymerization. In agreement with the behavior of the idealized system with microtubules of fixed length, approximate centering could be predicted under the conditions of stochastic length changes, as long as the microtubules were comparable in length with the radius of the confining boundary. Similarly, an eccentric position resulted, when the stochastically changing microtubules were significantly longer. In later numerical investigations that considered microtubule bending elasticity (Arkhipov and Maly 2006a; Howard 2006; Baratt et al. 2008; Kim and Maly 2009; Pinot et al. 2009; Letort et al. 2016), its effect was also compounded by other simultaneously acting mechanisms: not only the stochastic microtubule assembly, but cell surface dynamics, action of molecular motors, and hypothetical additional influences.

Equilibria of the confined microtubule cytoskeleton as dictated by microtubule bending elasticity alone have also been examined further (Maly and Maly 2010). To clarify the conditions for establishment of symmetry *vs.* asymmetry, the symmetric unstrained configuration was considered. Effects arising in flat and three-dimensional cytoplasmic domains were analyzed, in the geometries relevant to the experimentally well-studied cases of epithelioid cells and lymphocytes. The analysis demonstrated how the asymmetry of the microtubule cytoskeleton arises from instability of the symmetric equilibrium due to the confinement and microtubule bending elasticity. Since bending elasticity is a property of any microtubule cytoskeleton, these results imply that the observable asymmetries cannot be fundamentally attributed to the other factors that may be simultaneously at work, such as weak underlying asymmetry of the unstrained configuration of the cytoskeleton itself, external asymmetries, stochasticity, or other forces. Instead, asymmetry emerges inevitably in any confined microtubule cytoskeleton, insofar as its microtubules bend.

Stability analysis of the symmetric state led to the recognition that the asymmetric equilibrium in flat cells does not, in the general case, correspond to the absolute energy minimum. Instead, the asymmetric equilibrium displays special stability properties and history-dependence. This accounts for the different value of the coefficient of approximate proportionality between the equilibrium distance of the centrosome from the center and the difference of the microtubule length and cell radius: 1 (Maly and Maly 2010), compared with approximately 1.5 for the energy minimum (Holy 1997). In three-dimensional cells, however, the asymmetric equilibrium indeed corresponds to the absolute energy minimum. (The coefficient in three dimensions equals 2.) Also, the analysis established that while in two dimensions, the instability is third-order, in three dimensions it is of a special kind: infinitesimal displacements lead to a finite force. These and other results concerning the mechanics of the cytoskeleton confined by the cell boundary will be laid out below, following an exposition of how the problem can be formalized.

The analysis of centrosome positioning by microtubule bending can be approached by dividing the task of finding the equilibrium conformation of the cytoskeleton into, first, finding the equilibrium forms of microtubules, and then finding the equilibrium position of the centrosome (Maly and Maly 2010). Figure 6.1 shows the geometry related to the problem of finding the equilibrium of a single microtubule. Table 6.1 lists the model parameters. We will be concerned, in the main, with situations where microtubule length L exceeds cell radius R, and then

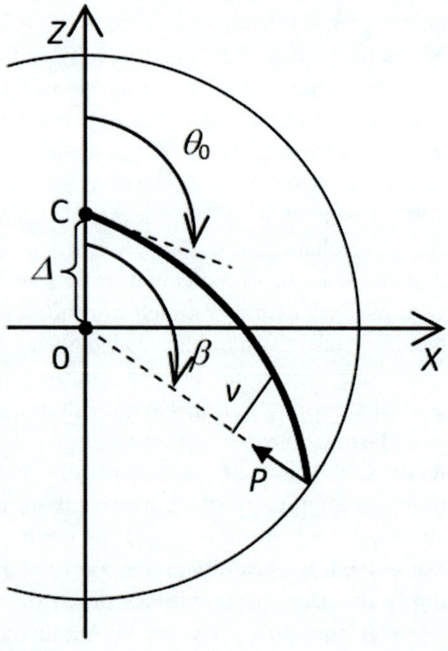

Fig. 6.1 Equilibrium of a confined centrosomal microtubule, with notation. *C*, centrosome. Reproduced from Maly and Maly (2010). Reprinted from Biophysical Journal, Vol. 99, V.I. Maly, I.V. Maly, Symmetry, Stability, and Reversibility Properties of Idealized Confined Microtubule Cytoskeletons, 2831–2840, copyright 2010, with permission from Elsevier

Table 6.1 Nomenclature of the symmetry instability model (Maly 2013). Reprinted by permission from Springer Nature Customer Service Centre GmbH: Springer Nature, Systems Biomechanics of the Cell by Ivan V. Maly (https://www.springer.com/gp/book/9781461468820), copyright 2013

Symbol	Meaning
β	Angular coordinate of the microtubule contact with the boundary
Δ	Distance between the centrosome and the cell center
Δ_e	Equilibrium distance between the centrosome and the cell center
θ	Angle of the tangent to the microtubule
θ_0	Angle at which the microtubule is clamped at the centrosome
EI	Microtubule flexural rigidity
F	Force exerted by all microtubules on the centrosome
f	Component of the force exerted by a microtubule on the centrosome along the axis of symmetry
L	Microtubule length
M	Moment of the microtubule-boundary contact force
N	Number of microtubules in the cell
n_i	Fraction of microtubules in stable (i = 1) and metastable (i = 2) forms
P	Force exerted by the cell boundary on the distal microtubule end
$p = N/(2\pi)$	Density of microtubules per unit of θ_0 in a flat cell
$p = N/(4\pi)$	Density of microtubules per unit of solid angle Ω in a three-dimensional cell
R	Cell radius
s	Axial coordinate along a microtubule
v	Moment arm of the microtubule-boundary contact force

with such microtubules as actually touch the cell boundary, but only with their distal ends. At the same time, we will always keep track of which of the microtubules touch the boundary in each cytoskeleton configuration, and which adopt a fully relaxed, straight form in the cytoplasm with their distal ends not in contact with the boundary.

Equations that specify bending of the single microtubule consist of the standard equilibrium beam equation and boundary conditions of clamping on the centrosome and frictionless contact with the cell boundary:

$$\frac{d}{ds}x(s)=\sin\theta(s),\quad \frac{d}{ds}z(s)=\cos\theta(s),\quad ds=\sqrt{dx^2+dz^2},$$

$$\frac{d}{ds}\theta(s)=\frac{M(s)}{EI},\quad M(s)=Pv(s),\quad v(s)=-x(s)\cos\beta+z(s)\sin\beta,$$

$$x(0)=0,\quad z(0)=\Delta,\quad \theta(0)=\theta_0,\quad x(L)=R\sin\beta,\quad z(L)=R\cos\beta.$$

Since

$$x(L)=x(L,\Delta,\theta_0,P,\beta),\qquad z(L)=z(L,\Delta,\theta_0,P,\beta),$$

the contact conditions, which are the characteristic equations for the nonlinear boundary problem, specify the unknown parameters P and β as functions

$$P = P(L,\Delta,\theta_0), \quad \beta = \beta(L,\Delta,\theta_0).$$

In the three-dimensional case, there is only one stable solution, and, correspondingly, one (P, β) pair. We will consider the physical reasons for this below. In the flat (two-dimensional) case, the nonlinearity leads to existence of two types of locally stable equilibrium forms of microtubules, and accordingly there are two branches of the functions for the unknown parameters:

$$P_i = P_i(L,\Delta,\theta_0), \quad \beta_i = \beta_i(L,\Delta,\theta_0), \quad i = 1,\ 2.$$

The nature of these alternative solutions, similarly, will be considered later. When the parameters (P, β) are known, the differential equations of bending specify the deformed shape of each microtubule and the stresses in it. In particular, the action of each microtubule on the centrosome can be determined. For each microtubule, we have the component of the force,

$$f(\theta_0) = -P(\theta_0)\cos\beta(\theta_0),$$

that contributes to the total force on the centrosome. In the two-dimensional case, the total force can be found as

$$F = \sum_{i=1}^{2}\int_0^{\pi} 2f(\theta_0)n_i(\theta_0)\,p\,d\theta_0,$$

and in the three-dimensional case, as

$$F = \int f(\theta_0)\,p\,d\Omega.$$

The equilibrium condition for the centrosome and for the microtubule cytoskeleton as a whole, is, of course,

$$F = 0.$$

Observe that when $\Delta = 0$, the equilibrium problem for the single microtubule is equivalent to the Euler problem about an elastic rod hinged on both ends. Indeed, even though the centrosomal end of the microtubule is not hinged, the moment of force P at the centrosomal end is zero, when $\Delta = 0$. The following approximation (Maly and Maly 2010) of the Euler-Lagrange solution to the Euler problem can be used:

$$P = \frac{\pi^2\kappa}{L^2}\frac{1+0.09\varepsilon^3}{1-\frac{1}{2}\varepsilon}, \quad \beta = \frac{2\sqrt{\varepsilon}}{1-\frac{5}{48}\varepsilon-\frac{1}{53}\varepsilon^2}.$$

Here, $\varepsilon \equiv (L\text{–}R)/L$. Owing to the accuracy of these approximations—0.25% and 0.05% respectively—the ballistic method can reliably refine the solution, and then be used to continue the solution with respect to Δ. The domain of existence of the second solution branch, which corresponds to metastable microtubule forms discussed below, in the two-dimensional case can be mapped numerically in the space (L, θ_0, Δ), and variable step sizes used to accurately account for the singularity at the boundary of this domain when continuing the solution and integrating the force.

The specific assumptions made in the above approach merit some discussion. The simple beam equation is the classical model for microtubule bending that has found both theoretical and experimental applications (see Howard 1998). The adequacy of the beam equation, which predicts the simple Euler forms of buckling, in application to intracellular microtubules is not universal in the light of observations of high-frequency buckling (Brangwynne et al. 2006). An example of more advanced models for bending of single microtubules can be found in the work of Gu et al. (2009). At the same time, the simple buckling forms compatible with the treatment laid out here evidently predominate in real cells. That the microtubules are clamped rather than hinged on the centrosome is similarly evident in living cells, for even strongly bent microtubules radiate from the centrosome in all directions before bending, instead of converging on it at sharp angles in a fan-like arrangement, which would be expected with hinging. It can be noted parenthetically that the situation concerning the pattern of convergence is the opposite in certain cell-free and centrosome-free experimental models *in vitro*.

Let us now consider the basic properties of the confined cytoskeleton, focusing first on the physics reflected in the multiplicity of solutions to the above equations. When the centrosome is in the center of the cell, the forms of a microtubule that differ in the direction of buckling are symmetric about the axis indicated by the orientation at which the microtubule is clamped (Fig. 6.2a). These forms, naturally, have the same energy. In three dimensions (in a spherical cell), an infinite number of such forms are connected by continuous rotation about the said axis. In two dimensions (in a flat round cell), there are only two such forms of a microtubule, and they correspond to diametrically opposed forms of the three-dimensional case.

With the displacement of the centrosome from the cell center, the equivalency of the microtubule's buckling forms is lost (Fig. 6.2b). Now, the form which is convex in the direction of the centrosome displacement is bent less, overall, and the opposite form is bent more. In three dimensions, these two diametrically opposed forms are the only remaining equilibrium forms. They lie in the plane defined by the displacement of the centrosome and by the unstrained direction of the microtubule. They are connected by a continuity of nonequilibrium forms, and there is no energy barrier between them. The microtubule therefore will adopt the lowest-energy form (the one which is convex in the direction of the centrosome displacement from the center). In two dimensions, the two equilibrium forms are connected only by higher-energy nonequilibrium forms in the plane of the cell. Therefore, both of the equilibrium forms are locally stable, and both can be occupied even after the

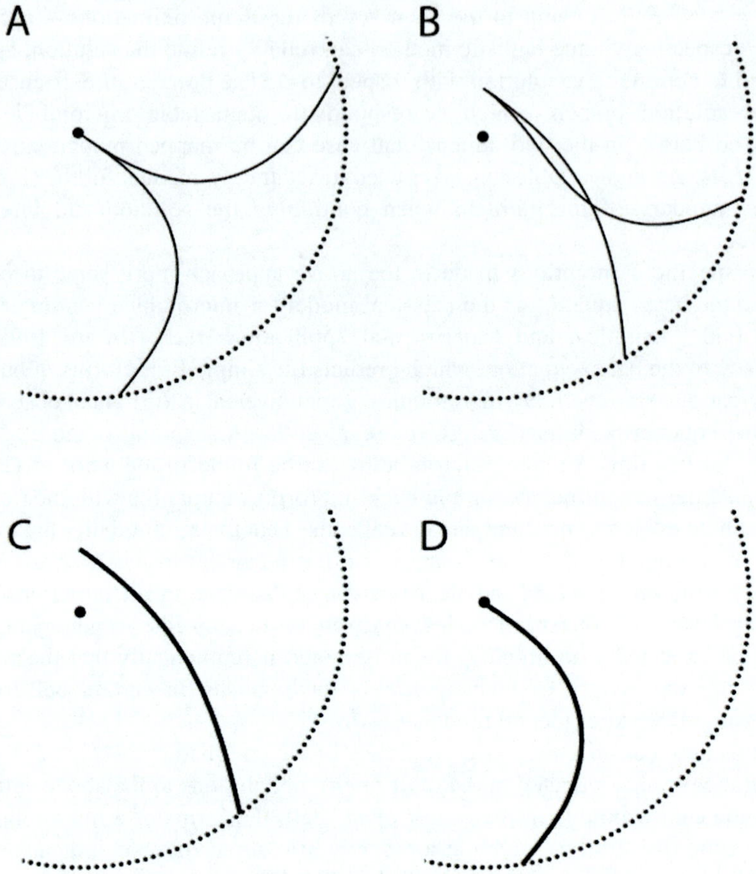

Fig. 6.2 Equilibrium forms of a microtubule. *Dot*, cell center. (**a**) Centrosome in the center, (**b**) Centrosome has been moved away from the center, (**c**) Further displacement of the centrosome (one form disappears), (**d**) Flat cell: centrosome has been moved back to the cell center from the last shown position; only one of the two potentially existing forms is occupied. Reproduced from Maly and Maly (2010). Reprinted from Biophysical Journal, Vol. 99, V.I. Maly, I.V. Maly, Symmetry, Stability, and Reversibility Properties of Idealized Confined Microtubule Cytoskeletons, 2831–2840, copyright 2010, with permission from Elsevier

displacement of the centrosome from the center. The higher-energy form can be called metastable, and the lower-energy form, stable.

As the deviation of the centrosome from the center increases, a position is eventually reached beyond which the metastable equilibrium form no longer exists. When this happens, the microtubules adopt the corresponding stable form (Fig. 6.2c). "Flipping" into the only remaining equilibrium is an energy-dissipating process, and cannot be reversed by the centrosome returning to the domain of existence of the metastable solution. This phenomenon underlies emergent irreversibility in an intrinsically elastic cytoskeleton, which is caused by its confinement within the cell boundary. We will consider the irreversibility effects in detail, once the simpler cases of aster equilibrium have been laid out.

The introduced equilibrium formalism is simplified by the residual symmetry that is preserved in the deforming, intrinsically symmetrical cytoskeleton. In principle, each equilibrium form of a bent microtubule is characterized by both force and torque that it exerts on the centrosome. The general condition for the equilibrium of the entire microtubule cytoskeleton inside the cell therefore stipulates that the total force as well as torque exerted on the centrosome by the microtubules are zero. However, the intrinsically symmetrical cytoskeleton is characterized not only by identical lengths of microtubules emanating from the centrosome in each direction, but also by a uniform distribution of the unstressed directions at which the microtubules are clamped at the centrosome. When this is the case, displacement of the centrosome from the center preserves one axis of symmetry, which coincides with the displacement. This residual symmetry ensures that the total torque is zero and the total force is collinear with the centrosome displacement.

These considerations prepare us for some elementary applications of the introduced formalism. The simplest example is provided by the spherical cell. In three dimensions, as has been mentioned, any deviation of the centrosome from the cell center specifies the direction of buckling for each microtubule, which is convex in the direction of the centrosome displacement. Thus, any such deviation generates a form of the cytoskeleton in which there is a non-zero total force on the centrosome, and the direction of this force is away from the center. The new equilibrium is reached when the centrosome is removed from the cell center by a distance that is approximately twice as large as the difference between the microtubule length and cell radius (Fig. 6.3). Comparison shows (Fig. 6.3) that the continuous treatment for the microtubule aster is already accurate when the number of microtubules is much lower than its typical value in mammalian cells. Figure 6.4 illustrates the shape of the three-dimensional cytoskeleton at equilibrium. The equilibrium conformation captures the salient features of the internal structure of nearly spherical real cells, such as lymphocytes (see, *e.g.*, Kuhn and Poenie 2002), namely the eccentric position of the centrosome and the "combed" arrangement of the microtubules.

So far, we have been assuming that all microtubules in the cell are of the same length. Generalization to a distribution of lengths is straightforward (Maly and Maly 2010). To preserve the intrinsic symmetry of the cytoskeleton, the distribution characterized by density function $q(L)$ should be the same for each orientation of unstressed emanation from the centrosome. The only modification to the derivations will be to integrate with respect to L in addition to integrating with respect to the emanation angle, when finding the total force F. The formula for the total force in the three-dimensional case becomes

$$F = \iint f\left(\theta_0, L\right) pq(L) d\Omega dL.$$

Let, for example, the distribution be uniform and cover the interval from $1.05R$ to $1.15R$. In this case, we obtain $\Delta_e = 0.220R$. In the situation with the constant length, when this length is equal to the mean of the distribution just considered ($L = 1.1R$),

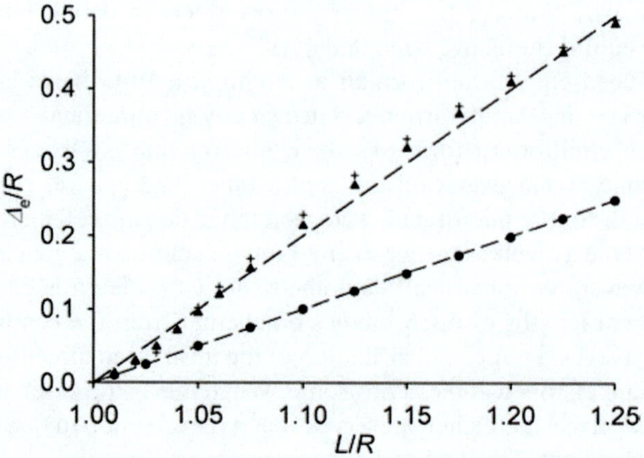

Fig. 6.3 Equilibrium distance of the centrosome from the cell center. *Triangles*, spherical cell in the continuous representation, *crosses*, spherical cell with 20 microtubules, *circles*, flat round cell in the continuous representation (positions reached spontaneously after a small perturbation of the fully symmetric cytoskeleton), *dashed lines,* slopes 1 and 2 for reference. Reproduced from Maly and Maly (2010). Reprinted from Biophysical Journal, Vol. 99, V.I. Maly, I.V. Maly, Symmetry, Stability, and Reversibility Properties of Idealized Confined Microtubule Cytoskeletons, 2831–2840, copyright 2010, with permission from Elsevier

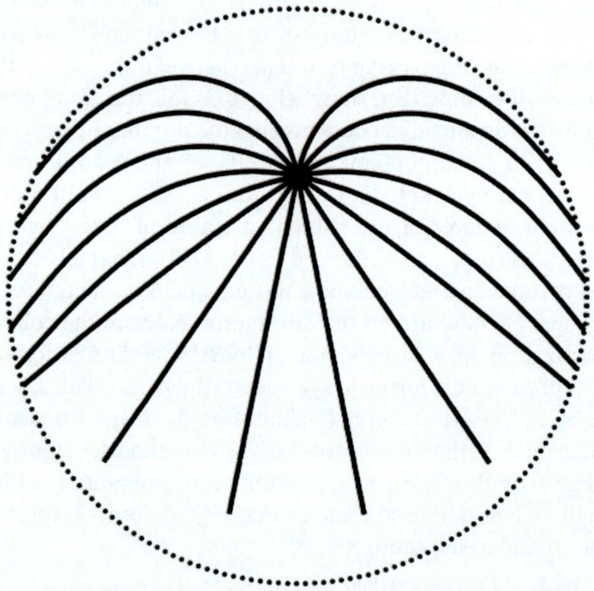

Fig. 6.4 Equilibrium conformation of the microtubule cytoskeleton in a spherical cell. Sample microtubule forms are shown that lie in the plane that is parallel to the plot and passes through the centrosome and cell center. Reproduced from Maly and Maly (2010). Reprinted from Biophysical Journal, Vol. 99, V.I. Maly, I.V. Maly, Symmetry, Stability, and Reversibility Properties of Idealized Confined Microtubule Cytoskeletons, 2831–2840, copyright 2010, with permission from Elsevier

the equilibrium distance is $0.214R$. We can see that the equilibrium distance with the distribution of lengths is not the same but larger, due to nonlinearity of microtubule bending.

The two-dimensional case corresponds to microtubules confined in an essentially flat spatial domain, as in the thinly spread cells that are cultured on glass in the typical experiment, and in various types of epithelial and similar cells *in vivo*. The lowest-energy conformation of the microtubule cytoskeleton, in which each microtubule is in the lower-energy (stable) equilibrium form, can be found the same way as in the three-dimensional case. It will share the "combed" appearance with the three-dimensional case (see Fig. 6.4), but this appearance is inconsistent with the cytoskeleton structure in flat cells, in which adjacent microtubules are, typically, buckled in both directions (see, *e.g.*, Euteneuer and Schliwa 1992). To understand the realistic two-dimensional structure, we need to consider the deformations leading to the equilibrium without assuming that the equilibrium actually reached corresponds to the global energy minimum among all possible conformations. This, in turn, requires specifying the initial conformation.

The full symmetry in the two-dimensional case is a reflection symmetry with respect to any axis that can be drawn through the center of the cell (Fig. 6.5a). This implies an infinite-fold rotational, or circular, symmetry, but excludes the case of vortex polarization that the rotational symmetry, by itself, would permit. The full symmetry thus requires that the centrosome be in the cell center, that the unstrained directions of microtubules be uniformly distributed around the centrosome, and that the two directions of buckling for each unstrained direction be equally represented. The fully symmetric conformation is, of course, a static equilibrium.

Small deviations of the centrosome from the center in the symmetrical two-dimensional cell result in a third-power growth of the total force on the centrosome (Maly and Maly 2010). This force is directed outward. As a result, the symmetry is unstable. A new equilibrium is reached, in which the centrosome is removed from the center by a distance approximately equal to the difference of the microtubule length and cell radius (Figs. 6.3, 6.5b, and 6.6). Because the equilibrium distance from the center is slightly smaller than the said difference, all microtubules in the flat equilibrium structure are in contact with the cell boundary and are bent, unlike in the three-dimensional case.

Irreversibility accompanies the spontaneous evolution of the symmetric flat microtubule aster in response to a perturbation of the centrosome position. The centrosome-displacing force that the microtubule system develops at small deviations of the centrosome from the center drives the system's evolution that is reversible insofar as the deformation remains small. However, metastable forms of the microtubules gradually become lost beginning with the first inflection point on the forward branch of the force function (solid curve in Fig. 6.6, point B). Beyond this point, a reversal of the outward movement of the centrosome places the system on a different branch of the force function. There is a separate reverse branch for each reversal point in the continuous treatment, which makes the force function infinite-valued. A system with a finite number of microtubules, naturally, has the number of reverse branches that is of the order of the microtubule number, *i.e.*,

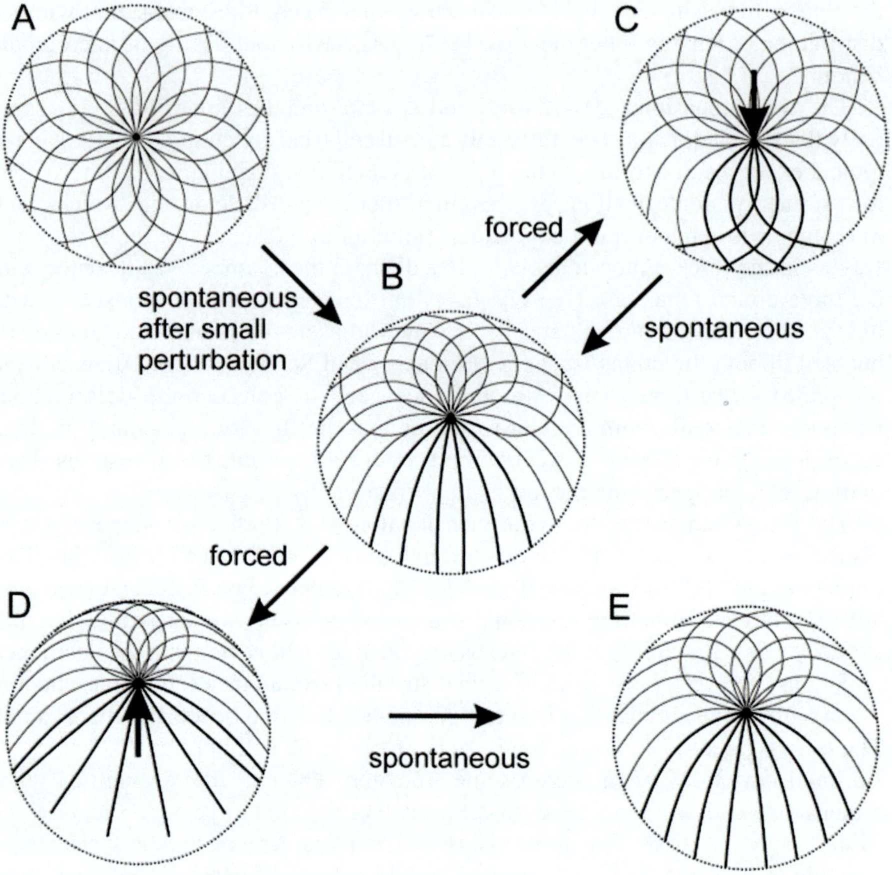

Fig. 6.5 Conformations of idealized microtubule cytoskeletons in a flat cell. (**a**) Fully symmetric cytoskeleton, (**b**) Equilibrium reached after a small perturbation, (**c**) Central centrosome position restored by an external force, (**d**) Additional deformation caused by an external force, (**e**) Equilibrium reached from the last configuration after removal of the external force. *Double-width lines* show forms that are doubly represented. Reproduced from Maly and Maly (2010). Reprinted from Biophysical Journal, Vol. 99, V.I. Maly, I.V. Maly, Symmetry, Stability, and Reversibility Properties of Idealized Confined Microtubule Cytoskeletons, 2831–2840, copyright 2010, with permission from Elsevier

hundreds in a typical flat cell maintained in the laboratory. In general, therefore, the force is a functional of the centrosome path.

As Fig. 6.6 illustrates, the reverse branches are non-zero at the zero centrosome displacement. Thus, although the central position of the centrosome can be restored by forces external to the microtubule cytoskeleton, the new central position will not be an equilibrium (Fig. 6.5c): The symmetry loss in response to a small perturbation proceeds spontaneously beyond the range of reversibility.

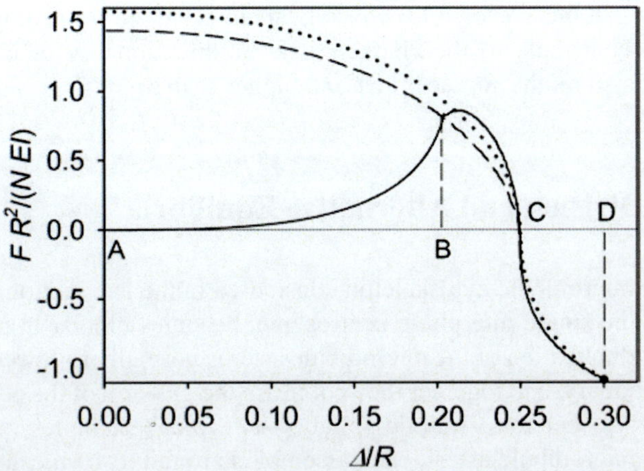

Fig. 6.6 Total force exerted by the microtubules on the centrosome as a function of the centrosome position in a flat cell. *Solid curve* is traced in the course of a continuous displacement of the centrosome away from the cell center, starting in the fully symmetric state. *Dashed* and *dotted curves* are traced during forced reversed displacement. Reproduced from Maly and Maly (2010). Reprinted from Biophysical Journal, Vol. 99, V.I. Maly, I.V. Maly, Symmetry, Stability, and Reversibility Properties of Idealized Confined Microtubule Cytoskeletons, 2831–2840, copyright 2010, with permission from Elsevier

One may also ask what happens if the centrosome is forced beyond the spontaneously achieved equilibrium (Fig. 6.5d). The reverse branches of the force function from beyond the equilibrium point pass very close, within 1%, of the original equilibrium (see the example in Fig. 6.6). Therefore, the irreversible effect of the perturbation is not detectable by recording the equilibrium centrosome position, and the eccentric equilibrium position of the centrosome in flat cells may, for practical purposes, be called stable. Even through the centrosome position is almost precisely restored after the removal of the external force, forcing the centrosome farther away from the center leaves an irreversible trace in the cytoskeleton structure (Fig. 6.5e). Taking this observation together with the irreversible loss of symmetry in response to infinitesimal perturbations, we can say that the equilibrium form of a microtubule cytoskeleton confined in a flat cell exhibits memory of past perturbations.

Irreversibility is the hallmark of life that has long been recognized from the perspective of biochemical thermodynamics. That energy derived from the nonequilibrium chemistry can also be dissipated in the mechanical manifestations of life, such as movement against viscosity and friction, is obvious. As the analysis of the confined microtubule aster demonstrates, additional, nontrivial irreversibility emerges specifically on the cellular level of structural organization. A different mechanism leading to the specifically cell-level, emergent irreversibility will be considered later in this chapter. The example already given, however, is sufficient to observe that the mechanical interaction of the cell's basic components—the interior organized by the microtubules and the cortex that forms the boundary—is generally irreversible. Tissues and developing organisms may have their own mechanisms leading to irreversibility, but the irreversible cytoskeletal effects in the constituting cells can be presumed to contribute to the characteristic unidirectionality of biological form development also on the supracellular and higher scales.

6.2 Cell Division and Alternative Equilibria

The unipolar microtubule cytoskeleton considered in the last section, which is centered around the single interphase centrosome, becomes bipolar in preparation for division. The duplication of the microtubule aster is normally followed by a division of the cell boundary, and together they comprise the division of the cell. The bipolar microtubule cytoskeleton—mitotic spindle—in typical examples consists, essentially, of two microtubule asters, each assembled around its own centrosome (spindle pole). Positioning of the mitotic spindle through the interaction of astral microtubules with the cell boundary often determines whether the cell division will be symmetric or asymmetric. This process plays a crucial role in the development of organisms and tissues, as well as in the segregation of the germ line.

Cells that divide symmetrically produce two daughter cells that are of equal or approximately equal size. Cell lines on which experiments are conducted in culture tend to reproduce in this fashion. Asymmetric divisions that result in daughter cells being of unequal size abound during development and differentiation in multicellular organisms. Experimentally well-characterized examples include single-cell embryos of the mussel *Unio* (Lillie 1901), roundworm *Caenorhabditis* (Hyman and White 1987) and leech *Helobdella* (Symes and Weisblat 1992), *Drosophila* neuroblasts (McCarthy and Goldstein 2006), mammalian oocytes (Schuh and Ellenberg 2008; Wei et al. 2020), and neuronal precursor cells (Delaunay et al. 2014). In the roundworm, for example, the first unequal division creates a larger cell that is the first somatic cell and a smaller cell that continues the germ line.

In addition to the significance of size as such—*e.g.*, between the large stem cell and its small progeny (Watt and Hogan 2000)—the division into daughter cells of unequal size may lead to an unequal distribution of specific components of the mother cell cytoplasm between the progeny. Such components may include cell fate determinants. In this connection, the general notion of asymmetric cell division includes cases where the daughter cells are of equal size, yet differ in the complement of components that they inherit (Sunchu and Cabernard 2020). Here, we will use the term "asymmetric division"exclusively in reference to the division that generates daughter cells of unequal size, which most obviously exemplifies the morphogenetic action of mechanical forces.

Usually, cells divide through the middle of the mitotic spindle (Bray 2001). The spindle proper consists of microtubule bundles that connect the two spindle poles. Precise terminology was developed (Maly 2012) in order to address the biomechanical question posed by the symmetric and asymmetric cell division through the positioning of the bipolar microtubule cytoskeleton within the still-common cell boundary. The line segment connecting the two spindle poles has been termed physical spindle axis. In the geometrical sense, this axis can be extended to define a coordinate axis that passes through the cell. The instance of the (extended) spindle axis passing through the cell center can be considered first. The paradigmatic cases, *e.g.*, HeLa cultured cells (Théry et al. 2005) or the first division in *Caenorhabditis* (Grill et al. 2001), exhibit this type of geometry. The cell center can be defined as the geometrical center of the space delimited by the cell boundary. The fundamental questions posed by the symmetric and asymmetric cell division are as follows: What determines coincidence of the center of the spindle proper with the cell center? What determines its shift from the cell center along the spindle axis?

Emanating from the spindle-pole centrosomes alongside the microtubules of the spindle proper, astral microtubules radiate to the cell periphery. As reviewed below (see also Pearson and Bloom 2004; di Pietro et al. 2016; Bergstralh et al. 2017), the spindle acquires its intracellular position through the astral microtubules' interaction with the cell boundary. Three types of effects are considered: the bending elasticity of microtubules that are deformed by their contact with the cell boundary, the stochastic assembly and disassembly of the microtubules, and pulling on the microtubules by molecular motors that are anchored on the cell boundary. The same hierarchy of contributions that we observed in the last section in connection with the unipolar microtubule cytoskeleton is operative here as well. The bending elasticity is an intrinsic property—all microtubules possess it. In contrast, the stochastic assembly and molecular motors may or may not play a role, depending on the specific intracellular conditions. Therefore, the latter two mechanisms always influence the spindle position in combination with the effects of bending elasticity. Understanding the effects of bending elasticity is, therefore, central to integrated biomechanical understanding of symmetric and asymmetric cell division.

The pioneering treatment by Bjerknes (1986) explained the salient features of spindle positioning in amphibian blastomeres. It dealt exclusively with bending elasticity, but microtubule deformations were not considered explicitly. It was assumed that straight astral microtubules radiate in all directions equally from the spindle poles. Each microtubule, by assumption, contacted the cell boundary with its distal end and developed the Euler buckling force that corresponded to its length. In addition to reproducing the orientations of spindles in the changing cell geometry of the progressively dividing embryo, it was demonstrated that the orientation exhibited bifurcations. Alternative equilibrium orientations appeared with the elongation of the spindle relative to the cell size. For intrinsically symmetrical spindles, however, only centered equilibria were considered, whereby the spindle's middle corresponded with the center of the cell. The adequacy of the straight-line and buckling-force approximations, and the stability of the equilibria to perturbations were not assessed.

Subsequently, in a model by Théry et al. (2005), the orientation of the spindle in a circular cell was treated as resulting from pulling on astral microtubules by force-generating elements anchored on the cell boundary. Only situations with central symmetry were considered. Microtubule deformations and bending elasticity were not included in the model. Force-balance models of pole separation and spindle morphogenesis (Cytrynbaum et al. 2003) treated the astral microtubules similarly.

Grill et al. (2001, 2005) analyzed the case of *Caenorhabditis* alternatively by neglecting bending or by approximating it with a Hookean dependence on the distance between the pole and the boundary. The equilibrium spindle position, as a result, either was centered or required an unequal distribution of cortical pulling elements to effect its shift from the centered conformation. Kozlowski et al. (2007) computed the microtubule deformations explicitly in three dimensions, but considered only such deformations as arising from dynamic drag in the cytoplasm, whereas the contact with the boundary was assumed to be "instantaneous" and not leading to any durable deformation resulting from confinement. It can be noted that in contrast with the reviewed simplifications of the treatment of the astral microtubules, deformations in the spindle proper (*i.e.*, of the microtubules that connect the two poles) have been treated with the precision of the standard bending elasticity theory (Nedelec 2002; Rubinstein et al. 2009).

Explicit consideration of large deformation of astral microtubules, caused by their confinement within the cell boundary, and stability analysis of the resulting equilibria (Maly 2012) permitted a general analysis of the effects of the microtubule elasticity on spindle positioning, as well as addressing the question of symmetric *vs.* asymmetric spindle positioning. Using these methods, it has been found that depending on the cellular parameters, the symmetric position of the spindle, as determined by microtubule bending alone, can be stable or unstable. Asymmetric stable equilibria are also characteristic of this minimal mechanical subsystem of the dividing cell, and two or more stable spindle positions can exist under the same parameters. This means that in experimental research aiming to elucidate the mechanisms and regulation of equal cell divisions, it is necessary to consider whether, in each individual case, the microtubule parameters are controlled by the cell so that the bending mechanics favors symmetry. If they are not, then it becomes necessary to ask what forces external to the microtubule cytoskeleton counteract the bending effects sufficiently to actively establish symmetry. Conversely, in the cases of unequal cell division, it is necessary to investigate whether the cell controls the microtubule parameters so that the bending favors asymmetry apart from any forces that are external to the microtubule cytoskeleton (see Delaunay et al. 2014).

As an extension of the formalism for the interphase, which was laid out in the last section, in the mitotic model we will be considering two asters around two rigidly coupled centrosomes at the spindle poles. These asters may be partial, in the sense that the microtubules may not radiate in all directions. Material spindle poles are small compared with the cell size, and the spindle proper, which connects them, contains a large mass of cross-linked microtubules (Bray 2001). The spindle proper, therefore, can be treated as an absolutely rigid segment that connects two points in space that represent the poles (centrosomes). In this respect, the model we are focusing on here (Maly 2012) is similar to the reviewed models by Bjerknes (1986), Kozlowski et al. (2007), and Théry et al. (2005). We will denote the interpolar distance, *i.e.*, the length of the spindle proper, S.

To address the problem as posed above, axially symmetric situations need to be considered, in which the axis of symmetry passes through the poles and, therefore, coincides with the spindle axis. In the Cartesian coordinate system that we are adopting, the x axis is collinear with the spindle axis. When isolated spindle poles are considered, and unless otherwise noted, the pole on the right will be meant. The pole coordinate is denoted x_p; the coordinate of the middle of the spindle proper is x_s. F_p will be the projection on the x axis of the total force acting on an isolated pole, and F_s, the projection on the x axis of the total force acting on the spindle (*i.e.*, the sum of the forces acting on each pole). F_p and F_s are collinear with the x axis due to the axial symmetry. Figure 6.7 illustrates the model and notation.

Microphotographs in the cited experimental work that was conducted on intact cells confirm that microtubules during mitosis do not converge on spindle poles at sharp angles in a fan-like arrangement despite bending against the cell surface and despite the poles' sometimes eccentric position. This rules out a hinged (free pivoting) boundary condition on the centrosome. The smooth bending forms seen in microphotographs are indicative of equilibrium flexure and absence of significant additional constraints. Accordingly, like with the microtubules in the interphase cell, we can treat the astral microtubules as rigidly clamped at the centrosomes, and their contact with the spherical cell surface, as frictionless. Since, as discussed in the last section, the alternative equilibrium forms of individual microtubules are unstable in three dimensions, we can focus on the ones with the lowest energy.

Unlike in the case of the interphase centrosome, which can be fully isotropic, in the spindle some microtubules that point inward from the spindle poles may not be astral but instead form a part of the spindle proper. Accordingly, the angle θ between the clamped direction and the outward direction of the spindle axis does not exceed a certain angle $\theta_{\max}$. We shall re-use notation N for the microtubule number and take

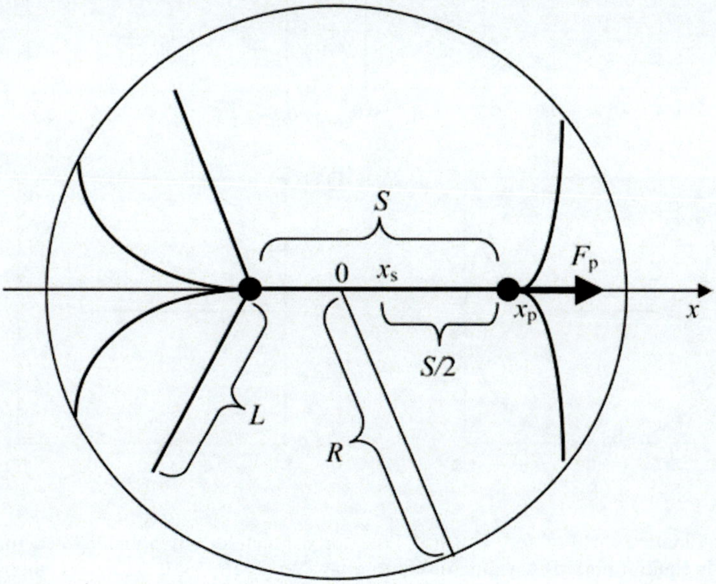

Fig. 6.7 Bipolar confined mitotic spindle. Schematic and notation. (Reproduced from Maly 2012 with permission)

it now to mean the number of the microtubules that radiate from one of the two poles. Intrinsically, the cytoskeleton structure that is of fundamental interest is centrally symmetric: *EI, L, N*, and θ_{max} are equal on both poles. Analysis shows (Maly 2012) that the behavior of this system, insofar as it concerns the problem of symmetric and asymmetric spindle positioning, is controlled entirely by three compound parameters: *L/R*, *S/R*, and θ_{max}. Moreover, the different system behavior regimes can be mapped in the space of these parameters, for practical purposes, exhaustively.

Limiting cases set the context for the mechanics of the system in the more complex, realistic regimes. The simplest behavior is exhibited by the greatly reduced structure in which astral microtubules emanate from the poles exclusively along the axis of the spindle ($\theta_{max} = 0$). As a pole moves collinearly with the spindle axis, the microtubules will abut on the cell boundary, buckle, and bend, exerting force on the pole. Sample conformations and force functions in this limiting case are presented in Fig. 6.8. When the axial distance of the pole from the boundary is greater than the length of the astral microtubules, the force is zero. When the distance is equal to the length, the magnitude of the force can take any value between zero and the buckling

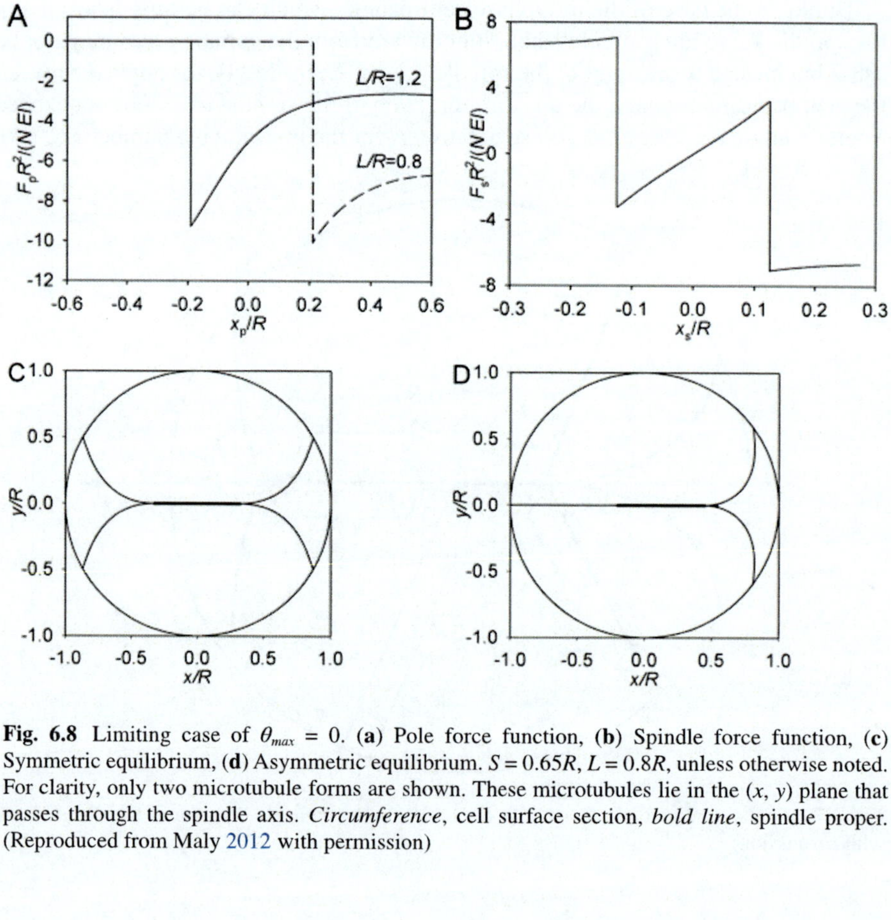

Fig. 6.8 Limiting case of $\theta_{max} = 0$. (**a**) Pole force function, (**b**) Spindle force function, (**c**) Symmetric equilibrium, (**d**) Asymmetric equilibrium. $S = 0.65R$, $L = 0.8R$, unless otherwise noted. For clarity, only two microtubule forms are shown. These microtubules lie in the (*x*, *y*) plane that passes through the spindle axis. *Circumference*, cell surface section, *bold line*, spindle proper. (Reproduced from Maly 2012 with permission)

force. For shorter distances, the force decreases, as the increasingly bent microtubules become less efficient at resisting the displacement of the pole. When the movement of two such poles is coupled through the spindle proper, three regimes are possible. If the spindle proper is short, the astral microtubules may not come in contact with the boundary. This happens when $L + S/2 < R$. The symmetric position will then be a neutral equilibrium. The special case of the entire structure just fitting in the cell without deformation is described by $L + S/2 = R$. Naturally, this rather special condition is unlikely to be realized, but it can be noted parenthetically that the models operating with microtubule buckling forces, such as the pioneering Bjerknes model, presuppose this condition while also implicitly assuming that it can be maintained as the centrosomes shift. The latter requirement is difficult to cast in physical terms even hypothetically. Returning to our limiting case, the non-trivial situation arises when $L + S/2$ is greater than R. In this case, microtubules emanating from one pole or from both must be bent.

Consider the symmetric case, in which both sides are bent equally. This, obviously, is an equilibrium. However, the magnitude of the force exerted on the pole in this situation is a locally decreasing function of the pole's distance from the boundary (Fig. 6.8a). Therefore, movement of both poles to the right will decrease the magnitude of the force exerted by the boundary on the right pole. By symmetry, the magnitude of the force exerted on the left pole will be increased. The symmetric equilibrium proves unstable. The coupled poles will continue moving spontaneously, until a new equilibrium is reached. In the second equilibrium, the force of the bent microtubules acting on one pole is balanced by the force of straight microtubules acting on the other pole. This is always possible, because the magnitude of the force of a bent microtubule is always lower than the buckling force. The asymmetric equilibrium is stable. Indeed, movement in the direction of the bent microtubules will leave the opposite pole unsupported, as the straight microtubules lose contact with the boundary. Movement in the opposite direction will place the system in a state it already passed during its spontaneous movement from the unstable symmetric equilibrium.

Thus, in this special case (the limiting case of $\theta_{max} = 0$), we observe instability of symmetry, stability of asymmetry, and determination of the stable conformation by the structural parameters that include S and L in relation to R. The forces are proportional to the microtubule bending rigidity, EI, and to the number of the microtubules, N, but the positions and stability of the equilibria do not depend on these parameters. The more general and realistic cases exhibit more complex behavior but retain these fundamental characteristics.

The opposite limiting case is also revealing—the special case of complete, intrinsically spherical asters at each pole ($\theta_{max} = \pi$). In this instance, the system behavior depends on whether the astral microtubules are longer or shorter than the cell radius. The case of short microtubules is simple. After the aster comes in contact with the boundary, the force exerted on the pole increases gradually with x_p. The smoothness of the force function is due to the number of the microtubules in contact with the boundary, which is increasing gradually, unlike in the case with $\theta_{max} = 0$. In addition, only the axial microtubules ($\theta = 0$), whose contribution to an aster with $\theta_{max} \neq 0$ is infinitesimal, go through developing the buckling force during the axial movement of the spindle; all other microtubules deflect on contact with the cell boundary. Bending leads to a decrease in stiffness, as can be seen in Fig. 6.9. The decrease in the stiffness of the aster (a progressively shallower slope of the force curve) is different from the previous case, where the magnitude of the elastic force was decreasing with the progressing deformation. The softening effect of the deformation in the case of the complete aster is more than offset by the increasing number of microtubules in contact with the boundary. Thus, even though the total force is a nonlinear function of the pole position, the force resisting the outward movement of the pole is monotonic (Fig. 6.9). Due to the monotonicity, the spindle with two complete asters of short microtubules exhibits stability of symmetry, and, in this respect, a behavior that is qualitatively similar to Hookean elasticity. In addition, the monotonicity means that there is only one equilibrium conformation of the mitotic microtubule cytoskeleton, insofar as the latter is large enough to maintain contact with the cell boundary.

Positioning of an isolated complete aster with comparatively long microtubules in three dimensions was considered in the previous section. In this case, now treated as a model for an isolated spindle pole, the pole force function has a root at approximately $x_p = 2(L–R)$: a more centrally positioned pole is attracted to the cell margin, whereas one positioned more eccentrically is repelled. Applying this result to a spindle that has an aster of this kind at each pole, we can observe that the presence of the root does not affect the stability of symmetry (Fig. 6.10). Even though the pole force function changes sign (Fig. 6.10a), it is monotonically decreasing for $x_p > 0$, like it did in the case of short microtubules. Thus, the symmetry of the spindle will be stable.

Overall, however, the pole force function with long microtubules is not monotonic (Fig. 6.10a). This opens the possibility of existence of additional equilibria, should both poles be on the same side of the cell. The system remains attracted to the symmetric equilibrium until the spindle proper is moved entirely into one half of the cell. Then, the conformational transition that was described in the last section in connection with the single three-dimensional aster occurs in the more central aster of the spindle. It places the mitotic system on the other branch of the spindle force function (Fig. 6.10b). The monotonicity of each half of the single pole force function ensures that following this transition, the spindle continues moving spontaneously in the direction that was previously forced. Indeed, even if the more distal pole is already in the region where it is repelled by the boundary, the more central pole upon crossing the center always experiences a greater force repelling it from the center. The new equilibrium is reached, when the two forces become equal in magnitude. This strongly asymmetric equilibrium is stable, due to the piecewise monotonicity: whether the poles are on different sides or on the same side of the cell, the force function (Fig. 6.10a) is locally decreasing for each pole. A centripetal movement of the spindle from the asymmetric equilibrium increases its repulsion from the center and decreases its repulsion from the nearest boundary, and vice versa. The spindle will therefore return to the asymmetric equilibrium, unless the

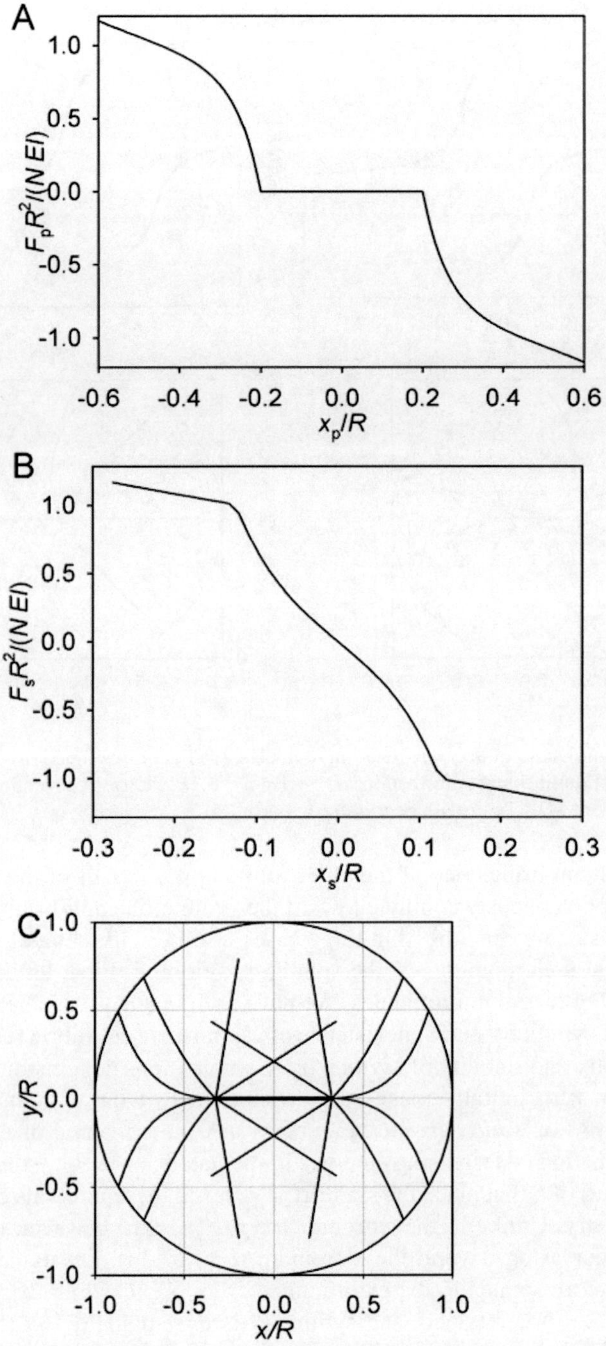

Fig. 6.9 Limiting case of $\theta_{max} = \pi$, short microtubules. (**a**) Pole force function, (**b**) Spindle force function, (**c**) Equilibrium conformation. $L = 0.8R$, $S = 0.65R$. For clarity, few microtubule forms are plotted. These microtubules lie in the (x, y) plane that passes through the spindle axis. Their values of θ are sampled uniformly between 0 and θ_{max}. (Reproduced from Maly 2012 with permission)

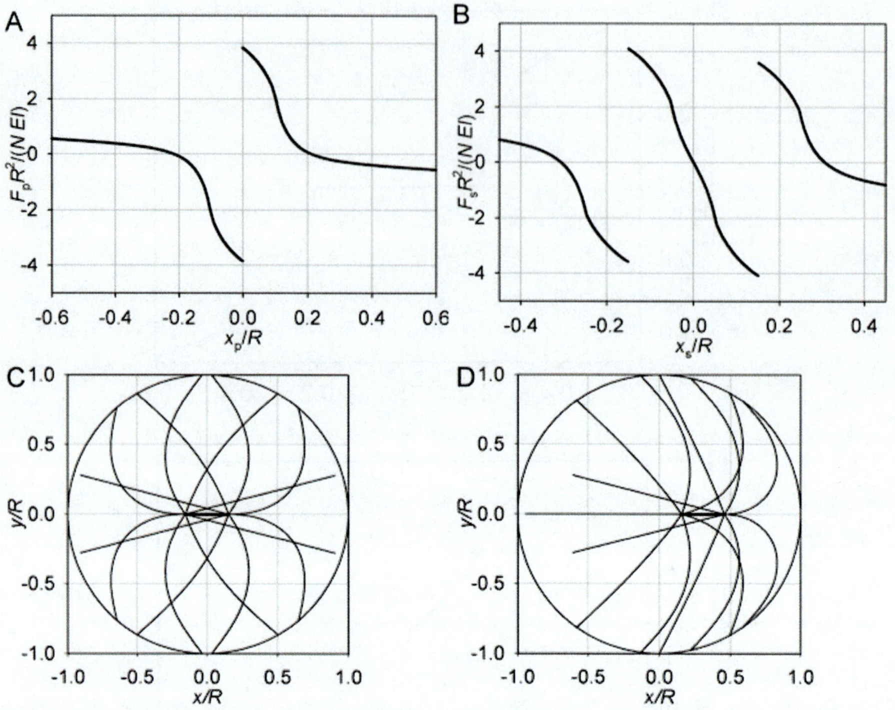

Fig. 6.10 Limiting case of $\theta_{max} = \pi$, long microtubules. (**a**) Pole force function, (**b**) Spindle force function, (**c**, **d**) Equilibrium conformations. $L = 1.1R$, $S = 0.3R$. Plotting conventions as in Fig. 6.9. (Reproduced from Maly 2012 with permission)

forced movement brings one of the poles to the opposite half of the cell, in which case it will spontaneously continue toward the symmetric equilibrium.

Transitions between the limiting cases further illuminate the systems-biomechanical underpinnings of the spindle positioning. First, let us consider the instance of short astral microtubules. We have seen that as θ_{max} increases from 0 to π, stability of symmetry and nonexistence of asymmetric equilibria replace instability of symmetry and stability of asymmetry. Computations demonstrate that increasing θ_{max} from zero initially makes the development of the extremum pole force smooth (Fig. 6.11a, *solid curve*, *cf.* Figure 6.8). A finite interval of x_p in which the pole force function is decreasing emerges immediately, once θ_{max} exceeds zero. The range of values of S that place the symmetric poles in these intervals corresponds to stable symmetry. Unlike in the extreme case of $\theta_{max} = \pi$, however, the pole force function is increasing beyond the extremum (Fig. 6.11a). Let us denote the pole position that corresponds to the extremum x_p'. For S exceeding $2x_p$,' the behavior seen with $\theta_{max} = 0$ is retained, and symmetry is unstable (Fig. 6.11b). The corresponding stable asymmetric equilibrium is derivative from the one described in the case of $\theta_{max} = 0$: One pole occupies a position that lies toward the cell center from the pole force function extremum, and the other occupies a position toward the cell

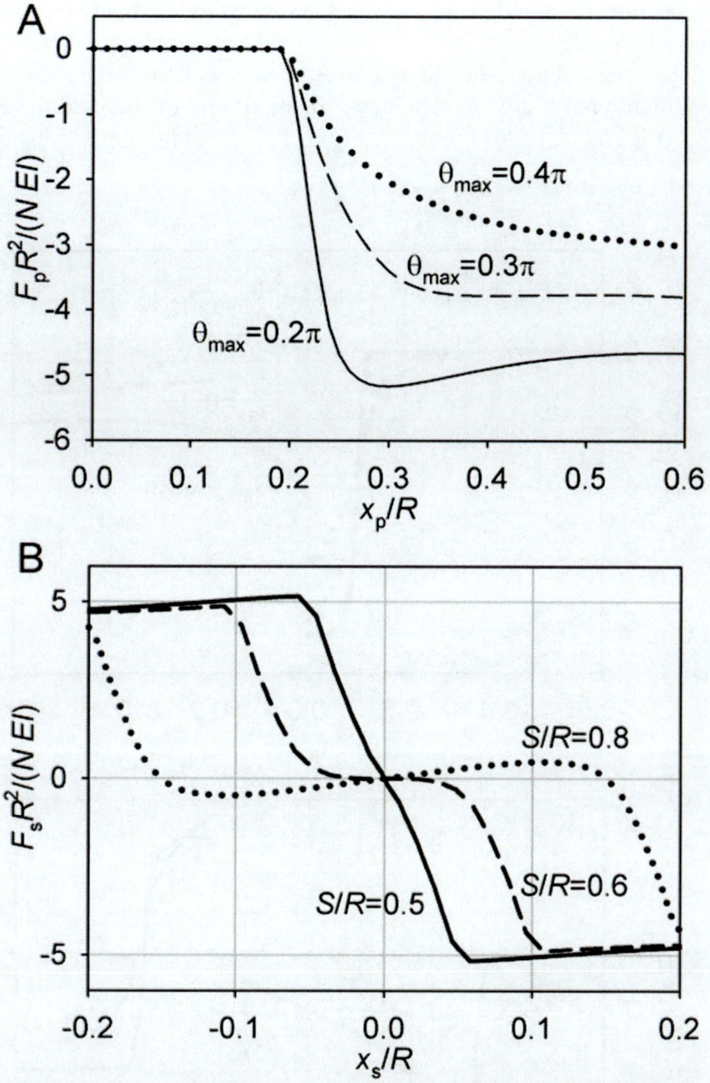

Fig. 6.11 Transition between θ_{max} = 0 and θ_{max} = π in the case of short astral microtubules. $L = 0.8R$. (**a**) Pole force function, (**b**) Spindle force function, θ_{max} = π/5. (Reproduced from Maly 2012 with permission)

margin from that pole's extremum. With a further growth of θ_{max}, the aster increases resistance at more eccentric pole positions (Fig. 6.11a, *dashed curve*). This eventually erases the nonmonotonicity of the pole force function. Figure 6.11a shows that for sufficiently large θ_{max}, the function becomes monotonic, like in the case of θ_{max} = π (*cf.* Figure 6.9). The spindle now exhibits only one equilibrium, which is symmetric and stable. Thus, with short astral microtubules, there is an abrupt transition at some intermediate θ_{max} to a behavior that qualitatively recapitulates the one observed with $\theta_{max} = \pi$.

With long microtubules, the increase of θ_{max} leads, at first, to the same changes as when the microtubules are short, namely to smoothing of the pole force function extremum (Fig. 6.12a, *solid curve*, *cf.* Fig. 6.11). In addition, the function in this

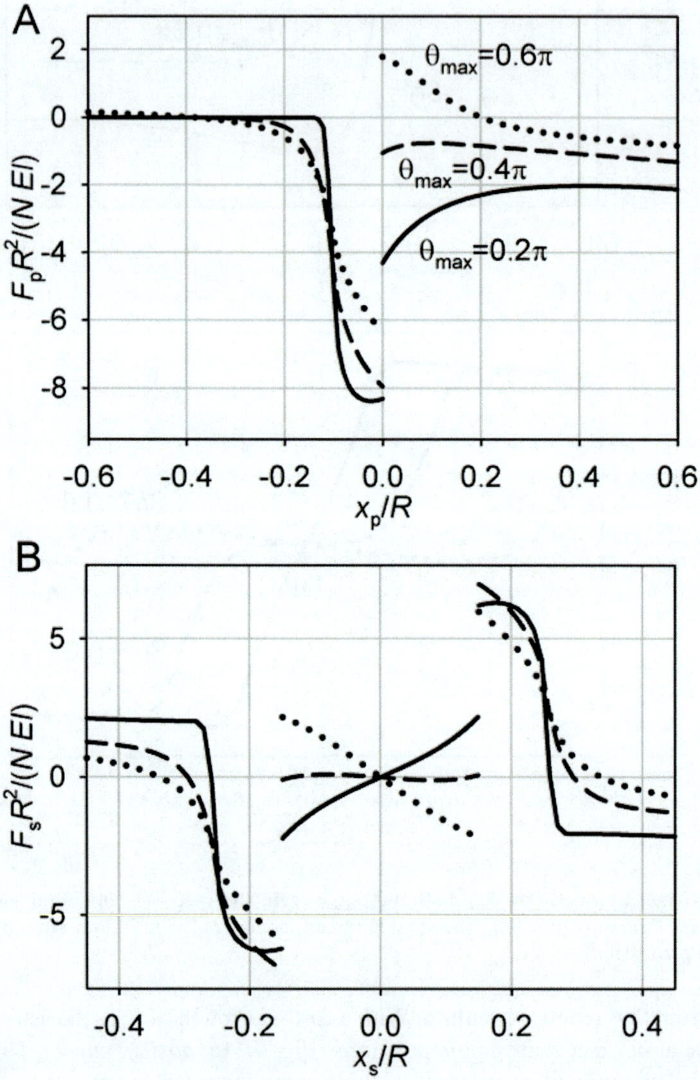

Fig. 6.12 Transition between $\theta_{max} = 0$ and $\theta_{max} = \pi$ in the case of long astral microtubules. $L = 1.1R$. (**a**) Pole force function, (**b**) Spindle force function. In (**b**), $S = 0.3R$, and the values of θ_{max} correspond to the line styles in (**a**). (Reproduced from Maly 2012 with permission)

case develops a discontinuity. It is caused by the conformational transition between the lowest-energy conformations, which differ on the two sides of $x_p = 0$, as was discussed in the previous section in connection with the interphase cytoskeleton. (The transition was non-observable in the case of $\theta_{max} = 0$, because the alternative conformations in that degenerate case were equivalent.) A more consequential difference from the case of the short microtubules is that the long microtubules are more deformable, and the drop of the force magnitude past the buckling-force maximum that they exhibit is deeper. Because of this, the increase in resistance at large pole displacements cannot erase the nonmonotonicity efficiently. Instead, at around $\theta_{max} = \pi/2$, the pole force function develops two descending branches (Fig. 6.12a). The difference from the already considered limiting case of complete asters (Fig. 6.10a) is that the pole force function with the intermediate θ_{max} may not change sign across the discontinuity. This is a reflection of the comparatively simple behavior of asters with $\theta_{max} < \pi/2$: the individual pole can find equilibrium, when all microtubules emanating from it are straight, unlike in more complete asters, where bending is unavoidable. Despite this difference in the mechanics of separate poles, poles coupled by the spindle behave in the same way as with the more complete asters of long microtubules, because of the fundamental similarity of the pole force function with two descending branches. The symmetric position is stable, as is the asymmetric equilibrium (Fig. 6.12b, *dashed curve*).

The intermediate case, $L \approx R$, presents special interest. Theoretically, it connects the qualitatively different types of behavior that have been described for long and short microtubules. Biologically, astral microtubules have length that is comparable with the cell radius in the morphogenetically important instances of large cells of early embryos (Lillie 1901; Hyman and White 1987; Symes and Weisblat 1992; Grill et al. 2001). By itself, of course, the case of $L = R$ is unrealistic, because the two quantities cannot be exactly equal. Notwithstanding, it can serve as a useful reference in the space of the spindle parameters and regimes of behavior.

Calculations show that a complete ($\theta_{max} = \pi$) aster with $L = R$ develops the peak force just before its pole reaches the cell center (Fig. 6.13a). This force is associated with the buckling force of the microtubules that all straighten when the pole reaches the center. When the pole is in the center, the elastic force can take any value between the positive and negative extremum, and will be zero for an aster not subjected to any external force. Thus, an individual separate aster exhibits a special kind of stability of the central position, in which a finite restoring force develops upon an infinitesimal perturbation.

However, the magnitude of the restoring force decreases with the magnitude of the perturbation (Fig. 6.13a). A pole that is closer to the center therefore experiences a higher centripetal force than a pole farther away from the center. This is a condition for instability of symmetry of the spindle. One pole's centripetal movement will be completed at the expense of the increasingly eccentric position of the coupled pole, as demonstrated by the spindle force function (Fig. 6.13b). In the asymmetric equilibrium, which is stable, the force on the eccentric pole is balanced by an equal force supported by the straight microtubules of the centrally positioned pole.

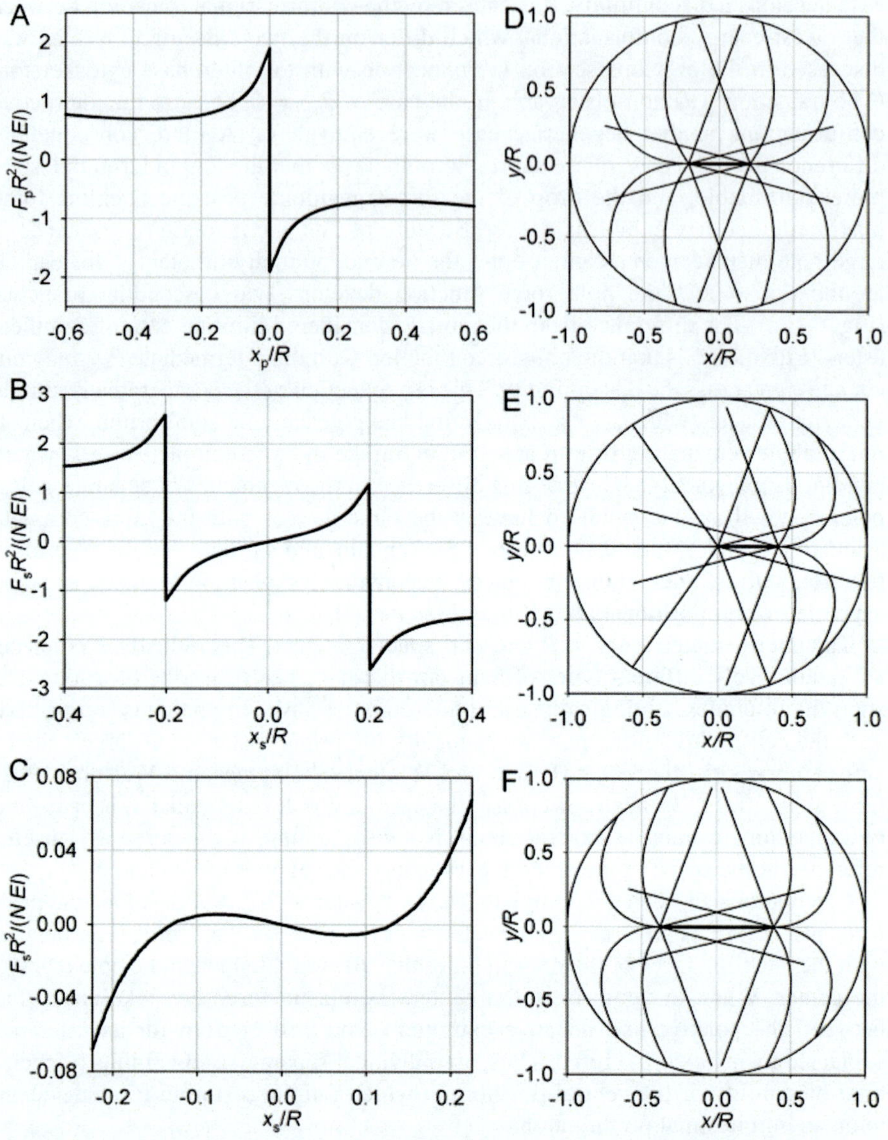

Fig. 6.13 Special case of $L = R$, $\theta_{max} = \pi$. (**a**) Pole force function, (**b**) Spindle force function, $S = 0.4R$, (**c**) Spindle force function, $S = 0.75R$, (**d**) Unstable symmetric equilibrium, $S = 0.4R$, (**e**) Stable asymmetric equilibrium, $S = 0.4R$, (**f**) Stable symmetric equilibrium, $S = 0.75R$. (Reproduced from Maly 2012 with permission)

The described behavior is observed for spindles with interpolar distances not exceeding a certain value. Inspection of the pole force function (Fig. 6.13a) shows that stiffening of the aster at large eccentricities of the pole creates a range of extreme pole positions where the magnitude of the force is increasing with the distance from the center. This creates conditions for stability of symmetry of sufficiently long spindles (Fig. 6.13c).

The domain of instability of spindle symmetry in the space of the structural parameters (θ_{max}, S/R, L/R) is outlined in Fig. 6.14. The behavior that we have explored in the foregoing examples can be seen in the shape of this domain, alongside some additional features. The boundaries of the domain show how the continuity between the regimes characteristic of the short and long microtubules is achieved through the special case of $L = R$. Specifically, for both long and short microtubules, the (θ_{max}, S/R) cross-sections of the symmetry instability domain are restricted to $\theta_{max} < \pi/2$.

For short microtubules, longer spindles exhibit unstable symmetry. Note, however, that the stability of spindles that are particularly short ($S < 2(R–L)$) is only neutral in this case, because in the symmetric position of such spindles, the astral microtubules do not contact the cell boundary. For long microtubules, instability of symmetry is exhibited by shorter spindles. In this case, it is of note that the longer spindles exhibit bistability between the symmetric and asymmetric equilibria, as was illustrated in Fig. 6.12b (*dashed curve*).

For $L \approx R$, the range of instability of symmetry extends through $\theta_{max} = \pi$. This behavior, which has been illustrated in the special case of $L = R$, is retained for L taking values between approximately 0.9 and 1.05R. Although this range is narrow in absolute terms, it is relevant to the morphogenetically important instances of large cells in early embryos (Lillie 1901; Hyman and White 1987; Symes and Weisblat 1992; Grill et al. 2001), where we find that L is indeed often similar to R.

It appears plausible that if each microtubule length from a set supports stability of symmetry, then any distribution of lengths over that set will also support stability of symmetry. Conversely, it may be posited that if each length by itself supports instability, any distribution will also support instability. Obviously, if any subset supports neutral stability, the outcome will be dictated by the complementary subset. With this in mind, predictions about the stability of symmetry with microtubule length distributions can be obtained by mere inspection of the stability and instability domains already mapped.

The instability domain in the (θ_{max}, S/R, L/R) space is bounded and embedded in the domain of stability (Fig. 6.14). For convenience, we can regard the L/R dimension as the vertical one. From this viewpoint, the instability domain lies above the domain of neutral stability, $S/R < 2\ L/R$. At large θ_{max}, there is a considerable separation between the domains of instability and neutral stability; at small θ_{max}, they almost touch. The support of a microtubule length distribution is represented in the described parameter space by a vertical segment. If such a segment lies entirely within or outside the instability domain, the central symmetry will be, correspondingly, unstable or stable. For example, any distributions whose support falls entirely to the right of the $L/R = 1$ isoline in Fig. 6.14b will specify stable central positioning of the spindle. Further, unless the distribution is unusually sharply concentrated at the intermediate values, the instability domain can be considered as touching the neutral stability domain for smaller θ_{max}. Therefore,

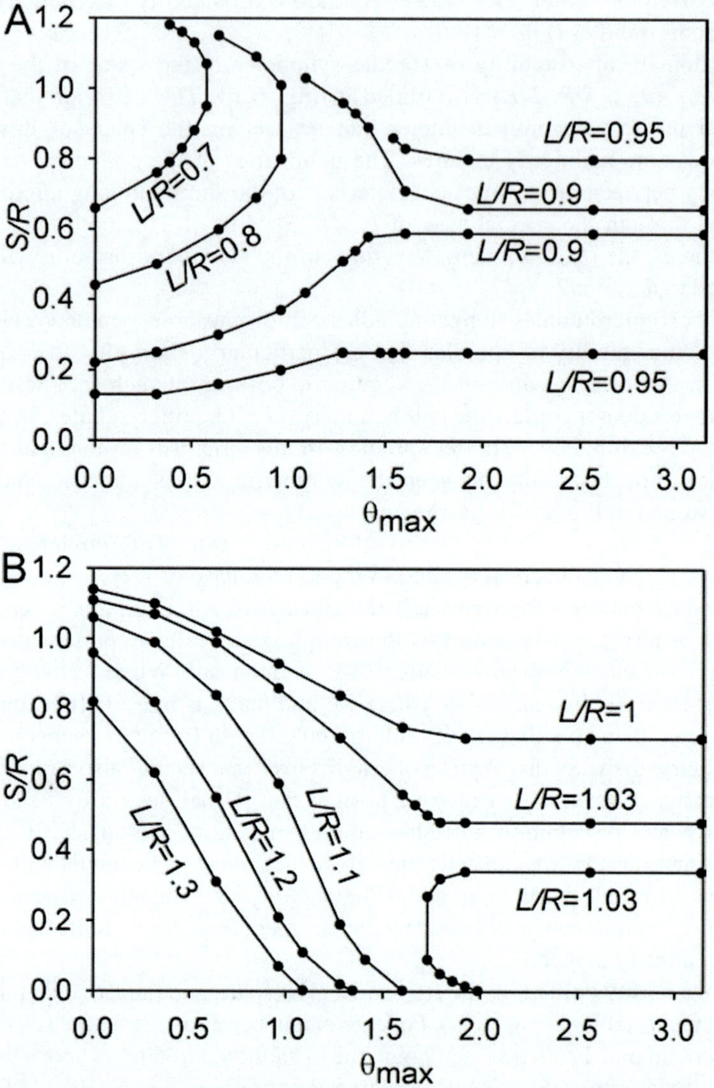

Fig. 6.14 Spindle symmetry instability domain. In the three-dimensional space of structural parameters (θ_{max}, *S/R*, *L/R*), *L/R* is understood as the vertical dimension in this graph. Isolines corresponding to the indicated values of *L/R* are shown. (**a**) Surface bounding the domain from below, (**b**) Surface bounding the domain from above. (Reproduced from Maly 2012 with permission)

distributions whose support falls below the upper bounding surface of the instability domain (Fig. 6.14b) specify instability of symmetry, if θ_{max} is not large.

In the common case of descending exponential distributions of microtubule lengths during mitosis (Gliksman et al. 1992), the contribution of the few exceptionally long microtubules may prove negligible. In that case, the specification of the outcome for spindle stability is further simplified: For larger θ_{max} and *S/R*, we expect stability, and for smaller θ_{max} and *S/R*, instability can be predicted.

Having set the context by considering the range of structures and equilibria that can be adopted by variously constructed (θ_{max}, *S/R*, *L/R*) mitotic spindles, let us now highlight several that are paradigmatic, based on the accumulated empirical information. Firstly, there is the case of a long spindle with short astral microtubules that radiate from the poles in a wide angle. The corresponding equilibrium conformation is plotted in Fig. 6.15a. According to the preceding analysis, in this regime (large *S/R*, small *L/R*, large θ_{max}), the symmetric equilibrium is the only equilibrium, and it is stable. This regime appears common among a variety of

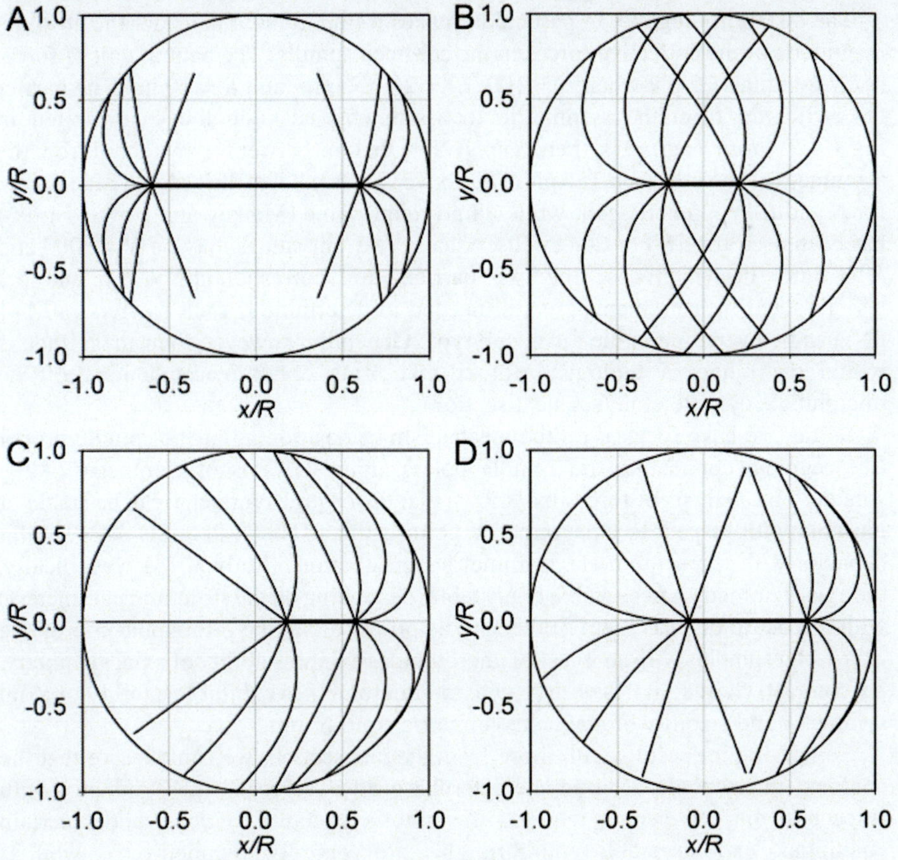

Fig. 6.15 Sample equilibrium conformations of mitotic spindles. Plotting conventions as in Fig. 6.9. In all examples, $\theta_{max} = 0.6\pi$. (**a**) $L = 0.7R$, $S = 1.2R$. (**b**, **c**) Alternative conformations that exist with $L = 1.1R$ and $S = 0.4R$. (**d**) $L = 0.95R$, $S = 0.7R$. (Reproduced from Maly 2012 with permission)

equally dividing cells. The HeLa cell line is one example that has been studied quantitatively (Théry et al. 2005).

Another characteristic example is the structure with long astral microtubules that radiate in a wide angle. This type of structure falls into the regime with bistability between the symmetric and asymmetric equilibria. Its alternative conformations are illustrated in Fig. 6.15b, c. Our example has a comparatively small S/R and large L/R. In its asymmetric equilibrium, it corresponds particularly to the first division in the invertebrate models of development that include the mussel *Unio* and the roundworm *Caenorhabditis* (Lillie 1901; Hyman and White 1987; Symes and Weisblat 1992; Grill et al. 2001).

For comparison, Fig. 6.15d demonstrates the only stable equilibrium—which is asymmetric—that exists in the regime with $L \approx R$ and a spindle proper of medium length. In terms of the spindle proportions and position, this last example describes the mouse oocyte (Schuh and Ellenberg 2008; Wei et al. 2020).

The bistability regime, in particular, makes it important to consider the absolute magnitude of the collective forces in the confined spindle. The natural unit of force, $N{\cdot}EI/R^2$, equals 22 pN when $N = 100$, $EI = 26$ pN μm², and $R = 10$μm. The barrier for switching from the symmetric to asymmetric state in the case plotted in Fig. 6.12b (*dotted curve*) is, then, 65 pN. Note that the barrier for switching from the asymmetric state is higher, 156 pN. The above values hold, to the order of magnitude, for N and R in a generic cell, while a mid-range value (Mickey and Howard 1995) has been used for EI. For other cells, equally relevant values may be $N = 1000$ and $R = 5$μm, in which case the two barriers are, correspondingly, 2.6 and 6.2 nN. Targeted measurements are still needed to determine if such barriers (~0.1–1 nN) can be overcome in the given cell type. Generally, however, their magnitude is within the range of biological forces that bring about rearrangement of the microtubule cytoskeleton (see next section).

So far, we have focused on the mechanism of positioning of the spindle proper (the segment connecting the spindle poles) along the axis of symmetry of the spindle. The analysis of forces has shown that the central symmetry can be stable or unstable with respect to axisymmetric perturbations. The system displays similar behavior with respect to nonaxisymmetric perturbations (Maly 2013). Specifically, the axial symmetry can be stable or unstable (depending on the structural parameters) with respect to displacement of the spindle proper along the orthogonal coordinate axis. Long spindles with short astral microtubules exhibit stability of axial symmetry. By contrast, short spindles with long astral microtubules exhibit instability of axial symmetry and stability of a nonaxisymmetric equilibrium.

Comparing the results with those from the last section, we can observe that the confined microtubule aster in the interphase always breaks the symmetry with respect to the cell center, whereas the mitotic structure, at least under certain conditions, exhibits stable equilibria that are centrally symmetric, as well as bistability between the symmetric and asymmetric equilibria. The formalism laid out here does not distinguish between equilibrium structures that are identical but for rotation with respect to the cell center, because microtubules in such structures are bent equally against the spherical cell outline. When the cell shape deviates from the sphere, the results should apply qualitatively, insofar as the position of the spindle along the common axis of the spindle and cell outline is concerned. An example of a structure where these conditions are relevant is the ellipsoidal egg of *Caenorhabditis*. In the other mentioned cases of the first embryonic divisions, in which the eggs are close to spherical, the formalism can be treated as an approximation.

Spontaneous development of asymmetry through bending of astral microtubules, and existence of special requirements for stability of symmetric conformations, have profound implications for designing and interpreting cell-biological and embryological experiments. Symmetric spindle positioning cannot be considered the "default" state of the system. Just as importantly, the source of the asymmetry cannot be sought exclusively outside an intrinsically symmetric structure consisting of the mitotic microtubule cytoskeleton and the confining cell boundary. When studying a case of symmetric positioning, it is worth investigating what makes it symmetric. Do the parameters such as the length of the astral microtubules (L) and of the spindle proper (S) take values that support the stable symmetry? If they do not, what forces external to the microtubule cytoskeleton act against the collective bending forces and actively establish the symmetry? Conversely, when studying a case of asymmetric spindle positioning, it is worth investigating the contribution of the collective bending effects of the astral microtubules to the generation of asymmetry. In fact, this question has priority. Although the quantitative nature of the collective bending that breaks symmetry is complex, the hypothesis that the source of asymmetry resides in the basic cytoskeleton structure is simple, compared with the alternatives of asymmetric regulation or asymmetric external forces, and should be put forward for testing in each particular case.

Similarly to the situation observed with the biochemical self-organization of cell polarity (Sect. 5.3) and the interphase cytoskeleton, spontaneous development of asymmetry through unequal bending in the astral microtubules does not by itself have a preferred direction: Each asymmetric equilibrium in the model has a counterpart, which is mirror-symmetric about the cell center. External forces and regulation mechanisms can, evidently, bias the spontaneous symmetry-breaking, even if they are not fundamentally needed for asymmetry generation. Action of the external forces, even a transient one, can also select between the symmetric and asymmetric equilibria in the cases of multistability. Forces arising from the cell's contact with other cells are fundamental in the hierarchy of such external influences, and will be considered next.

6.3 Cell Interaction and Multiperiodic Motion

Direct interaction with other cells can bias the spontaneous asymmetry in the given cell, or actively reposition a developed asymmetric intracellular structure into the physiologically functional orientation. A textbook example is the interaction of a T-killer lymphocyte with a tumor cell, or with a cell that has been infected and is producing viral particles (Bray 2001; Bykovskaja et al. 1978; Kupfer et al. 1982; Serrador et al. 1999; Gomez and Billadeau 2008; Sarkar et al. 2019; Chowdhury 2019; Hornak and Rieger 2020). In this crucial immunological mechanism, the T-killer cell patrols the tissue, comes in contact with an abnormal cell, recognizes a specific antigen on its surface, develops a firm contact with the target cell, and releases toxic compounds in its direction. The directionality of the release, which makes the killing efficient and spares the bystander cells, is as important as the precise molecular recognition of the antigen for the specificity of immune response (Kupfer and Singer 1989; Alberts et al. 2002). The killing apparatus in T cells is assembled around the centrosome at the center of convergence of the microtubules, making centrosome positioning the primary mechanism that brings about the immunologically functional orientation of the T cell structure.

As can be gleaned from the literature that has already been reviewed in connection with the asymmetry of the T lymphocyte structure (Sect. 6.1), the orientation of the centrosome in this system is under the control of a redundant set of cell-biological mechanisms, which is not surprising considering the importance of precision of the cellular immune response. However, the main mechanism (which is sufficient at least under the laboratory conditions) relies on cortical molecular motors. Specifically, dynein motors anchored at the T cell interface with the target "reel in" the centrosome by pulling on microtubules that pass over the interface (Kuhn and Poenie 2002; Combs et al. 2006). Large, complex oscillations of the centrosome next to the cell-cell interface follow the reorientation, and oscillations between the interfaces with simultaneously engaged targets are also observed. This dynamic situation is analogous (Maly 2011, 2013) to another model for microtubule cytoskeleton orientation, namely mitotic spindle positioning during the first divisions in *Caenorhabditis*, which lead to the segregation of the germ line. The specific formalism we will consider below has also been applied to a third experimental paradigm of cell polarity: the polarized epithelium with its characteristic cycles of centrosome and spindle positioning, which support oriented cell divisions and maintain the planar, polarized tissue (Kim 2018). The unipolar cytoskeleton structure in T lymphocytes and their interaction with one or two target cells being simpler, and their dynamics comparatively better understood, we will focus on this example.

The quantitative formalization of the reorienting T cell cytoskeleton (Kim and Maly 2009) rests on the following concepts. The cell boundary is a constant shape that consists of a round part and a flat part, the latter being attached to the target cell. The flat part is termed immunological synapse. Geometrically, it lies in the synaptic plane—a useful spatial reference for the microtubule and molecular motor arrangement. In the following analysis, we will define the angle between the synapse and the centrosome as the angle with the vertex at the center of the spherical segment of the T cell surface and the sides passing through the centrosome and the center of the synapse. Treating explicitly the dynamics of the microtubule cytoskeleton rather than being only concerned with stability of its equilibria, like we were in the preceding sections, requires taking into account the magnitude and distribution of the drag forces that limit its movement inside the cell. To this aim, we must consider mechanical coupling of the microtubule aster to the organellar complex associated with the centrosome, the largest part of which is the cell nucleus. Microtubules slide actively along the cell outline in the areas of contact with the immunological targets. As we already encountered in the last chapter, the directionality of the microtubule interaction with a specific type of molecular motor is determined by the directionality of the microtubule. In the present case, the motors being cortically anchored dyneins, the unit length of the part of the microtubule that is in contact with the synapse experiences a constant tangential force directed to the microtubule's distal end. The cell boundary in the synapse area is, therefore, characterized by a force density, and intrinsically, its action on the microtubules is isotropic. The magnitude of the force experienced by a microtubule depends linearly on the length of its segment in contact with the boundary, and the direction of the force is determined entirely by the orientation of the said segment.

In agreement with experiments on live cells, orientation of the centrosome toward the immunological synapse through the action of cortical dynein in the specified mechanical model proceeds to completion from many starting configurations (Kim and Maly 2009). Certain initial conditions for the cell-cell interactions (microtubule length, initial angle of contact with respect to the T cell centrosome) do not permit the complete reorientation by the synaptic dynein mechanism due to the constraints imposed by the confinement of the microtubule aster in the non-spherical boundary. In reality, these incomplete reorientations do not normally result in misdirected secretion of the active compounds, thanks to the complementary mechanisms, such as the previously mentioned whole-cell movement and selective dissolution of the misoriented cell-cell conjugate.

The initial condition of synapse formation at 90° to the centrosome has the highest likelihood, when the T cell comes in contact with the target cell in a random orientation. Successful reorientation over such a large angle results in an arrangement of the microtubules that is very asymmetrical. On the side of the microtubule aster that was leading in the reorientation movement, a relatively tight "bundle" of microtubules is formed, which is separated from the microtubules that were trailing by a distinctive gap (Fig. 6.16). "Bundling" of microtubules is frequently seen in models that do not involve any mechanism of the type of protein polymer cross-linking (Baratt et al. 2008; Kim and Rejniak 2014). The appearance of bundles arises from multiple microtubules adopting similar conformations that are favored by bending in the confined intracellular space. Under realistic intracellular conditions, of course, mechanical bundling can set the stage for a biochemical one. In the T cell-target cell conjugate, the gap between the leading "bundle" and the trailing microtubules has been successfully visualized by three-dimensional imaging (Kuhn and Poenie 2002). The strong asymmetry induced in the aster leads to a ratchet-like behavior of the microtubule cytoskeleton, when the T cell develops a second synapse. The centrosome readily reorients by another 90° in the same direction as it did the first time, but does not reorient in the opposite direction (Fig. 6.16). This adds a dynamic example to our exploration of irreversibility in the confined cytoskeleton, which has heretofore focused on quasi-static situations.

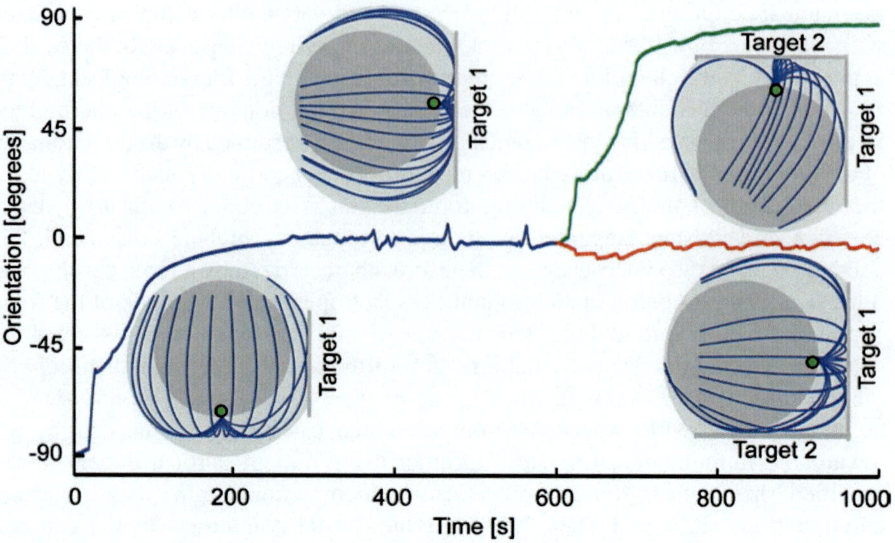

Fig. 6.16 Centrosome reorientation in a lymphocyte developing sequentially two immunological synapses. Insets show the model cell, nucleus, centrosome, microtubules, and cell contact areas. Two alternative simulations differing in the position of the second synapse follow the simulation with the first synapse. (Reproduced from Kim and Maly 2009 with permission)

The dynamic formalism reproduces emergence of pulse-like oscillations that follow the reorientation of the centrosome to the immunological synapse (Fig. 6.16). The oscillations are remarkable in the referenced live imaging data and have been linked to mechanical extrusion of the vesicles containing the active compounds at the target cell. Oscillations develop even if the synapse is formed next to the initial location of the centrosome. Engagement of the microtubules with the pulling surface causes the centrosome to move past (or away from) the center point of the interface. Although this point is the position of symmetry, complete symmetry does not develop, and cannot be sustained, in this system. The centrifugal movement of the centrosome eventually stops, and it begins reverse motion, again approaching the center point and again overshooting (Fig. 6.17). Oscillations persist indefinitely, without noticeable systematic changes, in the dynamic formalism. In real lymphocytes, of course, new conditions of cell-cell interaction develop, and the centrosome oscillations next to a given immunological synapse are not perpetual. Notwithstanding, they can be sustained in the laboratory observations over considerable time periods. Typically, as can be seen in Fig. 6.18, there are overlapping and interfering periodic motions. Also, oscillatory movements that are mostly tangential to the model cell-cell interface occur simultaneously with orthogonal oscillations. Gyrations (looping motions parallel to the interface) can also be discerned in the complex trajectory of the centrosome (Fig. 6.18).

Considering the mechanism of the oscillations and of the repeated overshooting exhibited by the centrosome, it is important to point out that inertia plays no role in

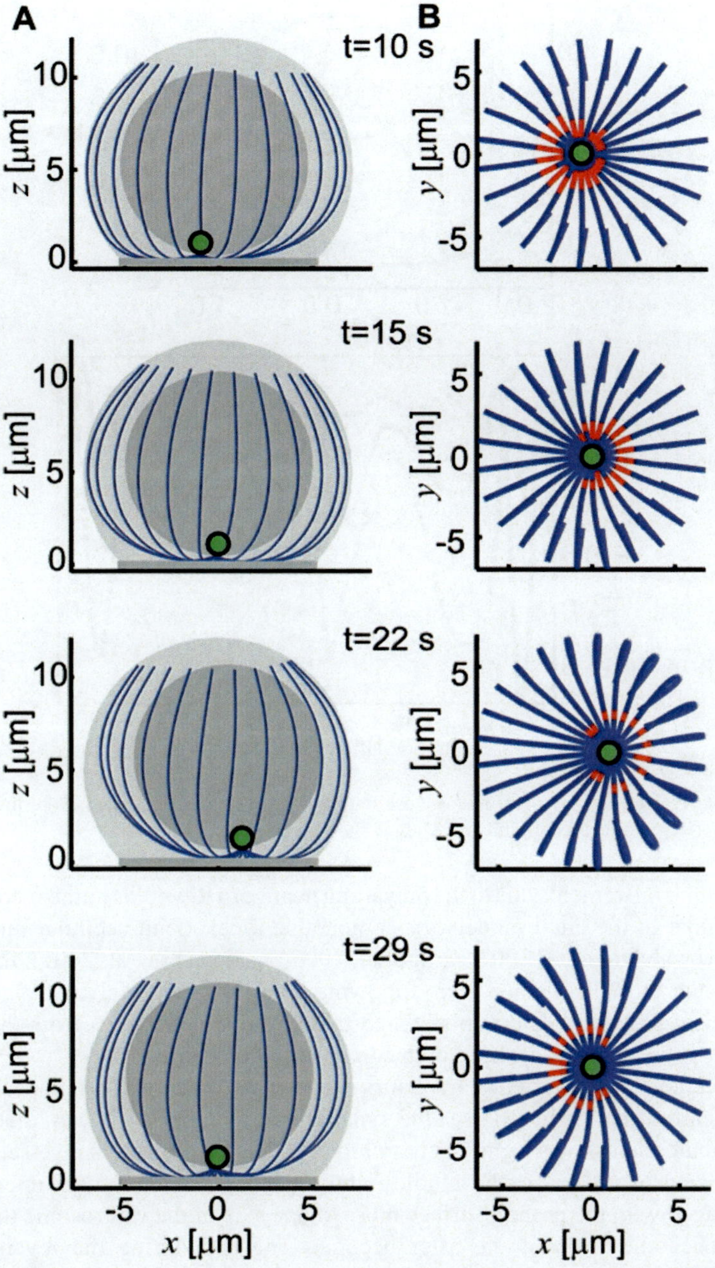

Fig. 6.17 Oscillations of the lymphocyte centrosome at the immunological synapse. (**a**) Cell structure plotted as in Fig. 6.16. (**b**) Microtubule system shown in projection onto the synaptic plane. The parts that are in contact with the synaptic surface and experiencing pulling are highlighted. (Reproduced from Kim and Maly 2009 with permission)

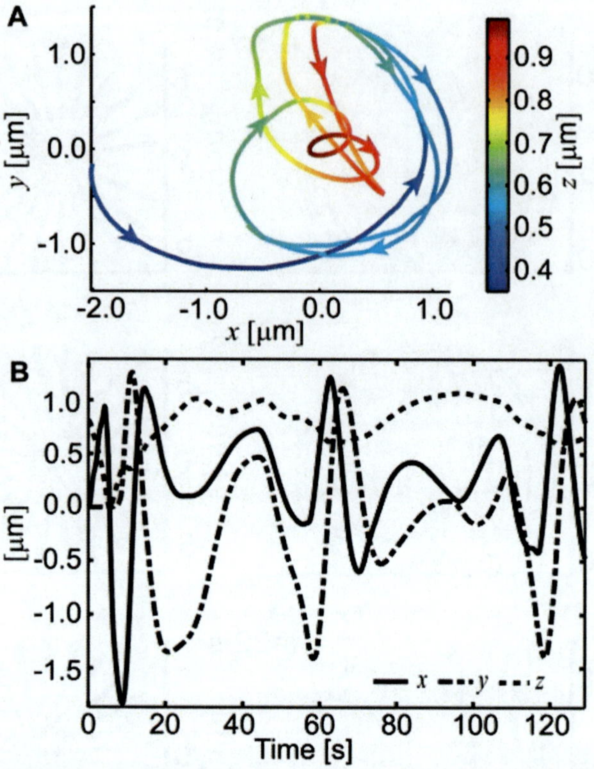

Fig. 6.18 Trajectories of centrosomes oscillating within a synaptic area. Coordinates as in Fig. 6.17. (Reproduced from Kim and Maly 2009 with permission)

intracellular movements due to the prevailing near-zero Reynolds number conditions. In fact, like in the cited models for comparable types of intracellular movements (*e.g.*, Cytrynbaum et al. 2003; Grill et al. 2005; Kozlowski et al. 2007), there is no mass in the formalism that we are following here (Kim and Maly 2009). Also, the mechanism is strictly deterministic, and therefore the deflections from the middle position of the centrosome are not due to molecular stochasticity.

Inspection of the system's trajectory through its conformational space reveals that when the centrosome passes the synapse middle point during oscillations, the microtubule aster shows significant asymmetry. This asymmetry is reversed the next time the centrosome passes the middle point (Fig. 6.17b). Moreover, the microtubules are engaged with the pulling surface more to one side of the centrosome than to the other. The other side of the aster becomes engaged during the reverse swing (Fig. 6.17b). We can also note that the distal ends of the microtubules do not move appreciably during the oscillation cycle. A similar condition has been observed in the Kozlowski et al. (2007) model for the pronucleus oscillations in worm eggs. It can be attributed to the cytoplasm viscosity dampening propagation of the elastic perturbation along the microtubules from their proximal parts, which are pulled by the cortex or coupled to the moving centrosome. As a result, when microtubules on one side are pulled and the centrosome shifts, the proximal parts of the microtubules on the opposite side become lifted off the synaptic surface (Fig. 6.17). This makes the tug of war nonlinear: when one side is winning, this weakens the opposing side. The temporal instability stems from this effect. Importantly for the cyclical nature of the movement that emerges from the instability, the range of motion is limited by the deformation of the microtubules on the winning side. Their distal parts are bent against the cell boundary, and because of this, the zone where they can contact the pulling surface does not extend to the edge of the flat synaptic zone. Movement toward the edge, therefore, diminishes the pulling force. This gives the elastic relaxation of the trailing microtubules time to catch up and bring these microtubules' proximal parts in apposition with the pulling surface. At this point, the microtubules that trailed are lying relatively flat on the synapse and are therefore experiencing a greater pulling force than the microtubules that led and are now contacting the synapse only with their highly curved parts. Movement in the reverse direction ensues (Fig. 6.17).

From the thermodynamic standpoint, the dissipative oscillations are driven by the non-potential forces of pulling. They arise from the energy-dissipating biochemistry of the dynein motors, which is coupled to the nonequilibrium metabolism of the living cell. In the course of the oscillations, these non-potential active forces work against similarly non-potential forces of viscous drag. At the same time, the mechanics of the described hysteretic loop is orchestrated by the potential forces of microtubule elasticity and by the structural constraints of the focused cytoskeleton and the cell boundary.

The basic frequency of the oscillations is fairly insensitive to the pulling force density. Yet, the pattern of the oscillations changes abruptly when this parameter crosses a certain value (Fig. 6.19). Below approximately 140 pN/μm, the oscillations appear as multiperiodic and continuous (Fig. 6.19a). Above 150 pN/μm, they are pulse-like (Fig. 6.19c). In the relatively narrow range, approximately 140 to 150 pN/μm, the oscillations are continuous and pure, *i.e.*, they exhibit a single frequency and amplitude. Only in this narrow intermediate range does the distance of the centrosome from the synaptic plane not oscillate (Fig. 6.19b).

The experimental estimates of the force exerted by a single cytoplasmic dynein molecule interacting with a microtubule, on the order of piconewtons (Ashkin et al. 1990), limit the range of realistic pulling force densities to 20–200 pN/μm. The period of oscillations parallel to the synapse in the simulation remains near 15–20 s throughout this range (Fig. 6.19d). This is close to the characteristic frequency recorded in the laboratory (Kuhn and Poenie 2002). While the accuracy of the approximation is a function of exact value of the effective viscosity of the cytoplasm, which is used with the dynamic formalism, the constancy of the period is an emergent property of the synapse pulling mechanism.

This intrinsic frequency of oscillations parallel to the synapse (x direction) is seen in its pure form when the orthogonal (z direction) oscillations are absent between 140 and 150 pN/μm (Fig. 6.19b). In the other two regimes (Fig. 6.19a, c), the dominant x-frequency is approximately the same (Fig. 6.19d). The period of the

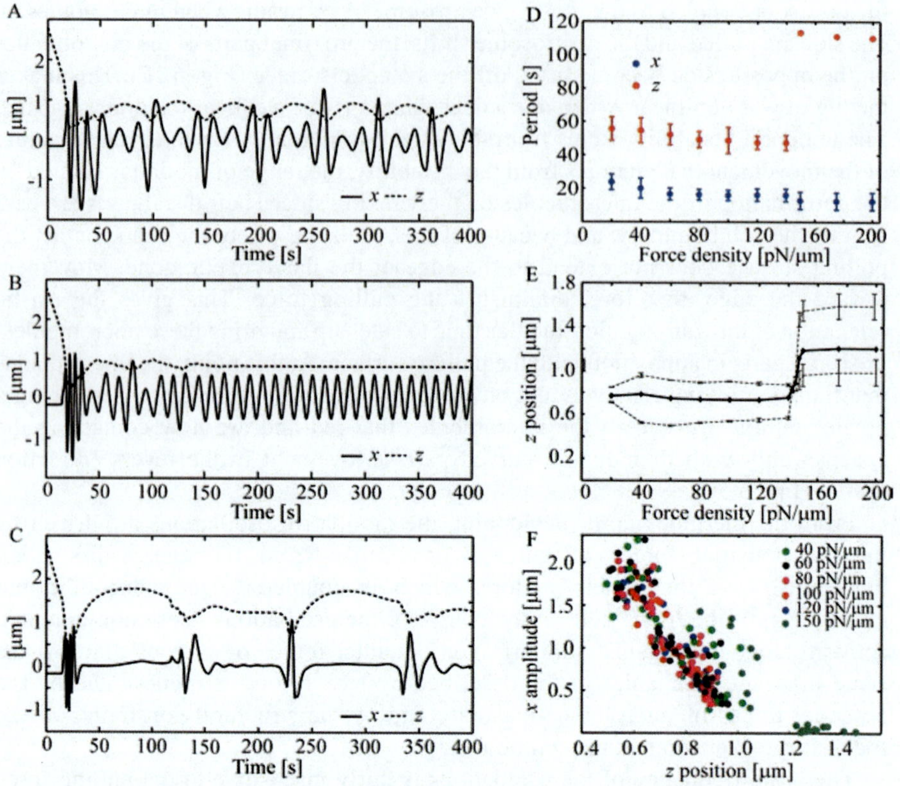

Fig. 6.19 Dependence of centrosome oscillations on pulling force density. (**a**) 100 pN/μm, (**b**) 143 pN/μm, (**c**) 200 pN/μm, (**d**) Period of oscillations that are parallel and perpendicular to the synapse, (**e**) Time average (*solid line*) and range (*dashed lines*) of the centrosome distance from the synapse, (**f**) Peak *x* deviation (amplitude) achieved at a given distance from the synapse, *z*. (Reproduced from Kim and Maly 2009 with permission)

z-oscillations is, similarly, mostly insensitive to the force density. However, it is much higher for all values above 150 pN/μm than below 140 pN/μm (Fig. 6.19d). In the intervening range, *z*-oscillations are not sustained (Fig. 6.19b).

The range of the distance (*z*) of the centrosome from the synapse exhibits a similar step-like dependence on the pulling force density, collapsing fully in the narrow transition zone (Fig. 6.19e). It can be observed that the farther away from the synapse the centrosome is at any given time, the smaller the amplitude of the movement parallel to the synapse will be. Figure 6.19f shows that this relation is quasi-linear and essentially independent of the force density. The exception to the linearity is brought about by the natural limit of zero amplitude. When this limit is reached, which happens only at high force density, the amplitude-distance relationship displays a breakpoint at the axis intercept (Fig. 6.19f). The zero amplitude of motion parallel to the synapse is observed during the intervals between the pulses, as shown in Fig. 6.19c.

Notably, the breakpoint of the *x*-amplitude *vs.* *z*-position curve (Fig. 6.19f) is near the centrosome-synapse distance of 1 μm, the same distance that characterizes the breakpoints of the dependencies of the *z*-position and *z*-period on force density (Fig. 6.19d, e). The said transition corresponds, in individual centrosome trajectories, to a complete but temporary loss of contact between the microtubule system as a whole and the synapse. This phenomenon arises from the viscous drag-induced "lift-off" of the microtubules (Fig. 6.17a). During particularly rapid movement that can occur at the higher force densities, not just one side but the entire microtubule system may lose contact with the synapse. In the absence of an active driving force, it takes the motile system a considerable time to relax elastically against the cytoplasmic drag and contact the synapse again. These time periods correspond to long, high arcs of the *z*-trajectory and no *x*-movement, as seen in Fig. 6.19c.

Complete characterization of the movement complexity that arises in this relatively simply constructed mechanical system is a difficult task. Notwithstanding, the numerical analysis demonstrates how the multi-periodic and variable-amplitude movement seen in the laboratory emerges from the biological force-generating mechanism at the cell-cell interface.

Following a simultaneous development of two synaptic areas (which can be set up experimentally and presumably occurs also in real tissues), the centrosome in the mechanical model moves to one of them. After pausing at the first synapse, which pause can last for a significant period of time, the centrosome moves spontaneously to the other synapse (Fig. 6.20a). The cycle of movement, pause, and movement to the other synapse continues indefinitely with a well-defined periodicity. The same considerations limiting the duration of such movement in real cell-cell interactions as we already had to consider in connection with oscillatory movement next to one synapse, apply here as well. At the same time, the characteristic delay before the reverse movement matches the centrosome behavior seen in the laboratory (Kuhn and Poenie 2002). In the numerical model, the angle between the two synaptic planes must be narrower than 150° for the delay to take place (Fig. 6.20b), which is consistent with the angle observed in the experiments. By adjusting the pulling force density and effective cytoplasm viscosity, both the duration of the pause and the duration of the movement phase can be reproduced (Fig. 6.20b). Whereas the exactness of the match of the absolute model time scale to the experiment is not particularly significant, the fact that the computed phases of pause and migration have the same relative duration additionally validates the mechanical model.

Analysis of the dynamic formalism shows that the characteristic pause of the centrosome and the associated secretory apparatus in the lymphocyte next to each of the synaptic areas results from delayed relaxation of microtubules that were trailing during the last period of centrosome migration. This can be discerned in Fig. 6.20, and it is the same factor that leads, in the extreme, to the irreversible, ratchet-like behavior associated with especially long-range reorientations (Fig. 6.16). In comparison, the medium-range migration between two nearby synapses can be reversible. On the other hand, if we compare it with the relatively small-amplitude oscillations within a synapse (Fig. 6.17), the migration of the centrosome between the synapses winds the trailing microtubules around the

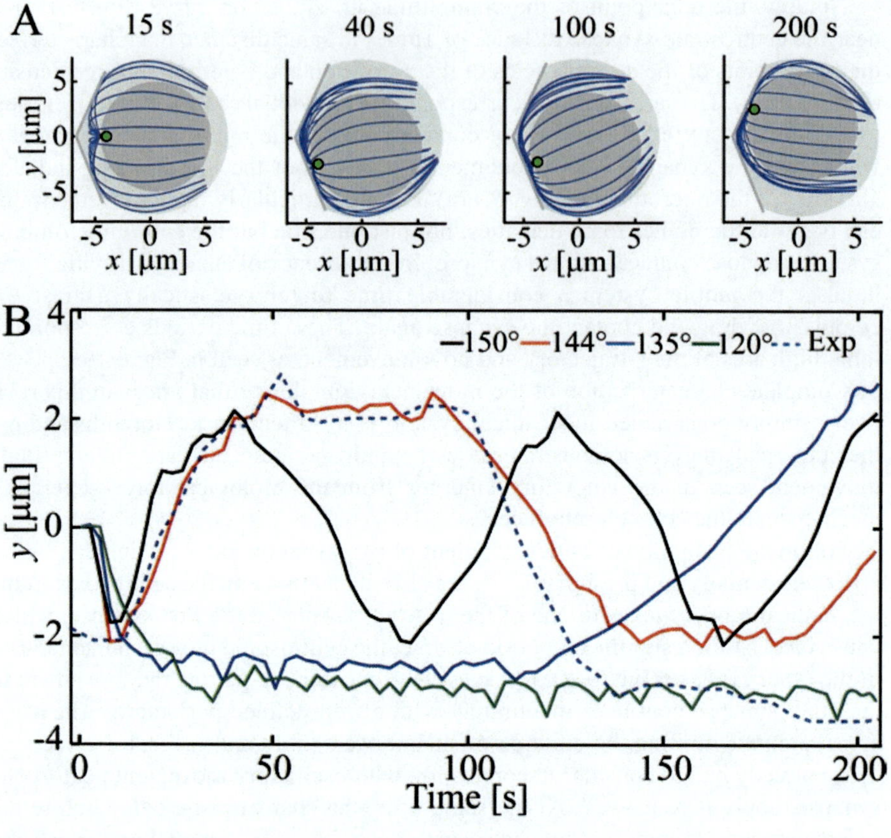

Fig. 6.20 Oscillations of the lymphocyte centrosome between two immunological synapses. **(a)** Model cell structure, **(b)** Trajectories of the centrosome calculated for the indicated angles between the synaptic planes. *Dashed curve*, experimental trajectory. (Reproduced from Kim and Maly 2009 with permission)

nucleus more. It takes them longer to relax and contact the second synaptic area, after the movement has been limited by deformation of the leading microtubules. It is notable that the time that the secretory apparatus associated with the lymphocyte centrosome spends next to the given target is determined by the elasticity of the cytoskeleton, while static equilibrium is never achieved during the apparent pauses.

In the preceding sections of this chapter, we have considered, firstly, mechanical instability, which leads to adoption of alternative equilibria and emergence of directional internal cell structure. We have also considered irreversibility, which imparts directionality to this structure's development. As shown here on some salient and better-studied examples, forces arising from cell contact work on the results of the asymmetry-generation to achieve *specific* directionality of the cell structure—one that is functional in the tissue context. On this level as well, the mechanics of the confined cytoskeleton can make the directionality acquisition irreversible. Additionally, the forces associated with cell-cell interaction create conditions for temporal instability. Dissipative oscillations that ensue are structurally hysteretic due to effects that are closely related to the ones at play during the irreversible directionality acquisition. The cyclical nature of these movements may be transparent in reduced mechanical models; their characteristic multiperiodicity, however, is difficult to distinguish from non-periodic movement in real systems. Therefore, more generally, it is preferable to speak of the relatively simple cellular forces' capacity to generate complex continuous motion. Although these aspects of intracellular dynamics appear fundamentally similar in the reviewed cases across the widely different taxa and cell types, much work remains to fully characterize the emergent effects of cellular forces. There is little doubt that further studies in systems biomechanics of the cell will close the gap between our understanding of the composition and self-organization on the macromolecular level, on the one hand, and cell or tissue structure on the other. Organ- and organism-scale morphogenesis that results from collective behavior of large numbers of cells will be the subject of the next chapter.

References

Abrahamsen G, Sundvold-Gjerstad V, Habtamu M et al (2018) Polarity of CD4+ T cells towards the antigen presenting cell is regulated by the Lck adapter TSAd. Sci Rep 8:13319

Alberts B, Johnson A, Lewis J et al (2002) Molecular biology of the cell. Garland Science, New York

Arkhipov SN, Maly IV (2006a) Contribution of whole-cell optimization via cell body rolling to polarization of T cells. Phys Biol 3:209–219

Arkhipov SN, Maly IV (2006b) Quantitative analysis of the role of receptor recycling in T cell polarization. Biophys J 91:4306–4316

Arkhipov SN, Maly IV (2007) A model for the interplay of receptor recycling and receptor-mediated contact in T cells. PLoS One 2:e633

Arkhipov SN, Maly IV (2008) Retractile processes in T lymphocyte orientation on a stimulatory substrate: morphology and dynamics. Phys Biol 5:16006

Ashkin A, Schütze K, Dziedzic JM et al (1990) Force generation of organelle transport measured in vivo by an infrared laser trap. Nature 348:346–348

Baratt A, Arkhipov SN, Maly IV (2008) An experimental and computational study of effects of microtubule stabilization on T-cell polarity. PLoS One 3:e3861

Bergstralh DT, Dawney NS, St Johnston D (2017) Spindle orientation: a question of complex positioning. Development 144:1137–1145

Bjerknes M (1986) Physical theory of the orientation of astral mitotic spindles. Science 234:1413–1416

Brangwynne CP, MacKintosh FC, Kumar S et al (2006) Microtubules can bear enhanced compressive loads in living cells because of lateral reinforcement. J Cell Biol 173:733–741

Bray D (2001) Cell movements: from molecules to motility. Garland, New York

Burakov A, Nadezhdina E, Slepchenko B et al (2003) Centrosome positioning in interphase cells. J Cell Biol 162:963–969

Bykovskaja SN, Rytenko AN, Rauschenbach MO et al (1978) Ultrastructural alteration of cytolytic T lymphocytes following their interaction with target cells: I. hypertrophy and change of orientation of the Golgi apparatus. Cell Immunol 40:164–174

Chowdhury D (2019) Laying tracks for poison delivery to "kiss of death": search for immune synapse by microtubules. Biophys J 116:2057–2059

Combs J, Kim SJ, Tan S et al (2006) Recruitment of dynein to the Jurkat immunological synapse. Proc Natl Acad Sci U S A 103:14883–14888

Cytrynbaum EN, Scholey JM, Mogilner A (2003) A force balance model of early spindle pole separation in *Drosophila* embryos. Biophys J 84:757–769

Delaunay D, Cortay V, Patti D et al (2014) Mitotic spindle asymmetry: a Wnt/PCP-regulated mechanism generating asymmetrical division in cortical precursors. Cell Rep 6:400–414

Di Pietro F, Echard A, Morin X (2016) Regulation of mitotic spindle orientation: an integrated view. EMBO Rep 17:1106–1130

Dupin I, Camand E, Etienne-Manneville S (2009) Classical cadherins control nucleus and centrosome position and cell polarity. J Cell Biol 185:779–786

Etienne-Manneville S, Hall A (2001) Integrin-mediated activation of Cdc42 controls cell polarity in migrating astrocytes through PKCzeta. Cell 106:489–498

Euteneuer U, Schliwa M (1992) Mechanism of centrosome positioning during the wound response in BSC-1 cells. J Cell Biol 116:1157–1166

Faivre-Moskalenko C, Dogterom M (2002) Dynamics of microtubule asters in microfabricated chambers: the role of catastrophes. Proc Natl Acad Sci U S A 99:16788–16793

Gliksman NR, Parsons SF, Salmon ED (1992) Okadaic acid induces interphase to mitotic-like microtubule dynamic instability by inactivating rescue. J Cell Biol 119:1271–1276

Gomes ER, Jani S, Gundersen GG (2005) Nuclear movement regulated by Cdc42, MRCK, myosin, and actin flow establishes MTOC polarization in migrating cells. Cell 121:451–463

Gomez TS, Billadeau DD (2008) T cell activation and the cytoskeleton: you can't have one without the other. Adv Immunol 97:1–64

Gotlieb AI, May LM, Subrahmanyan L et al (1981) Distribution of microtubule organizing centers in migrating sheets of endothelial cells. J Cell Biol 91:589–594

Grill SW, Gönczy P, Stelzer EH et al (2001) Polarity controls forces governing asymmetric spindle positioning in the *Caenorhabditis elegans* embryo. Nature 409:630–633

Grill SW, Kruse K, Jülicher F (2005) Theory of mitotic spindle oscillations. Phys Rev Lett 94:108104

Gu B, Mai YW, Ru CQ (2009) Mechanics of microtubules modeled as orthotropic elastic shells with transverse shearing. Acta Mech 207:195–209

Harold FM (2001) The way of the cell. Oxford University Press, New York

Holy TE (1997) Physical aspects of the assembly and function of microtubules. Dissertation, Princeton University

Holy TE, Dogterom M, Yurke B et al (1997) Assembly and positioning of microtubule asters in microfabricated chambers. Proc Natl Acad Sci U S A 94:6228–6231

Hornak I, Rieger H (2020) Stochastic model of T cell repolarization during target elimination I. Biophys J 118:1733–1748

Howard J (1998) Mechanics of motor proteins and the cytoskeleton. Sinauer, Cambridge

Howard J (2006) Elastic and damping forces generated by confined arrays of dynamic microtubules. Phys Biol 3:54–66

Hyman AA, White JG (1987) Determination of cell division axes in the early embryogenesis of *Caenorhabditis elegans*. J Cell Biol 105:2123–2135

Kim M (2018) Mechanism of MDCK II cell polarization during the cell division: a computational study. Appl Math Comput 317:1–11

Kim M (2019) A numerical mechanical model integrating actin treadmilling and receptor recycling to explain selective disengagement of immune cells. Math Biosci 316:108244

Kim MJ, Maly IV (2009) Deterministic mechanical model of T-killer cell polarization reproduces the wandering of aim between simultaneously engaged targets. PLoS Comput Biol 5:e1000260

Kim MJ, Rejniak KA (2014) Mechanical aspects of microtubule bundling in taxane-treated circulating tumor cells. Biophys J 107:1236–1246

Koonce MP, Tikhonenko I (2018) Centrosome positioning in *Dictyostelium*: moving beyond microtubule tip dynamics. Cell 7:29

Koonce MP, Köhler J, Neujahr R et al (1999) Dynein motor regulation stabilizes interphase microtubule arrays and determines centrosome position. EMBO J 18:6786–6792

Kozlowski C, Srayko M, Nedelec F (2007) Cortical microtubule contacts position the spindle in *C. elegans* embryos. Cell 129:499–510

Kuhn JR, Poenie M (2002) Dynamic polarization of the microtubule cytoskeleton during CTL-mediated killing. Immunity 16:111–121

Kupfer A, Singer SJ (1989) Cell biology of cytotoxic and helper T cell functions: immunofluorescence microscopic studies of single cells and cell couples. Annu Rev Immunol 7:309–337

Kupfer A, Louvard D, Singer SJ (1982) Polarization of the Golgi apparatus and the microtubule-organizing center in cultured fibroblasts at the edge of an experimental wound. Proc Natl Acad Sci U S A 79:2603–2607

Kupfer H, Monks CKF, Kupfer A (1994) Small splenic B cells that bind to antigen-specific T helper (Th) cells and face the site of cytokine production in the Th cells selectively proliferate: immunofluorescence microscopic studies of Th-B antigen-presenting cell interactions. J Exp Med 179:1507–1515

Letort G, Nedelec F, Blanchoin L et al (2016) Centrosome centering and decentering by microtubule network rearrangement. Mol Biol Cell 27:2833–2843

Levy JR, Holzbaur ELF (2008) Dynein drives nuclear rotation during forward progression of motile fibroblasts. J Cell Sci 121:3187–3195

Lillie F (1901) The organization of the egg of *Unio*, based on a study of its maturation, fertilization, and cleavage. J Morphol 17:227–292

Lowin-Kropf B, Smith Shapiro V, Weiss A (1998) Cytoskeletal polarization of T cells is regulated by an immunoreceptor tyrosine-based activation motif-dependent mechanism. J Cell Biol 140:861–871

Maly IV (2011) Systems biomechanics of centrosome positioning: a conserved complexity. Commun Integr Biol 4:230–235

Maly IV (2012) Equilibria of idealized confined astral microtubules and coupled spindle poles. PLoS One 7:e38921

Maly IV (2013) Systems biomechanics of the cell. Springer, Boston

Maly VI, Maly IV (2010) Symmetry, stability, and reversibility properties of idealized confined microtubule cytoskeletons. Biophys J 99:2831–2840

McCarthy EK, Goldstein B (2006) Asymmetric spindle positioning. Curr Opin Cell Biol 18:79–85

Mickey B, Howard J (1995) Rigidity of microtubules is increased by stabilizing agents. J Cell Biol 130:909–917

Nedelec F (2002) Computer simulations reveal motor properties generating stable anti-parallel microtubule interactions. J Cell Biol 158:1005–1015

Pearson CG, Bloom K (2004) Dynamic microtubules lead the way for spindle positioning. Nat Rev Mol Cell Biol 5:481–492

Piel M, Meyer P, Khodjakov A et al (2000) The respective contributions of the mother and daughter centrioles to centrosome activity and behavior in vertebrate cells. J Cell Biol 149:317–329

Pinot M, Chesnel F, Kubiak JZ et al (2009) Effects of confinement on the self-organization of microtubules and motors. Curr Biol 19:954–960

Rubinstein B, Larripa K, Sommi P et al (2009) The elasticity of motor-microtubule bundles and shape of the mitotic spindle. Phys Biol 6:016005

Sarkar A, Rieger H, Paul R (2019) Search and capture efficiency of dynamic microtubules for centrosome relocation during IS formation. Biophys J 116:2079–2091

Schuh M, Ellenberg J (2008) A new model for asymmetric spindle positioning in mouse oocytes. Curr Biol 18:1986–1992

Serrador JM, Nieto M, Sánchez-Madrid F (1999) Cytoskeletal rearrangement during migration and activation of T lymphocytes. Trends Cell Biol 9:228–233

Stowers L, Yelon D, Berg LJ et al (1995) Regulation of the polarization of T cells toward antigen-presenting cells by Ras-related GTPase CDC42. Proc Natl Acad Sci U S A 92:5027–5031

Sunchu B, Cabernard C (2020) Principles and mechanisms of asymmetric cell division. Development 147:dev167650

Symes K, Weisblat DA (1992) An investigation of the specification of unequal cleavages in leech embryos. Dev Biol 150:203–218

Tang N, Marshall WF (2012) Centrosome positioning in vertebrate development. J Cell Sci 125:4951–4961

Théry M, Racine V, Pépin A et al (2005) The extracellular matrix guides the orientation of the cell division axis. Nat Cell Biol 7:947–953

Ueda M, Gräf R, MacWilliams HK et al (1997) Centrosome positioning and directionality of cell movements. Proc Natl Acad Sci U S A 94:9674–9678

Watt FM, Hogan BL (2000) Out of Eden: stem cells and their niches. Science 287:1427–1430

Wei Z, Greaney J, Loh WN et al (2020) Nampt-mediated spindle sizing secures a post-anaphase increase in spindle speed required for extreme asymmetry. Nat Commun 11:3393

Yvon AMC, Walker JW, Danowski B et al (2002) Centrosome reorientation in wound-edge cells is cell type specific. Mol Biol Cell 13:1871–1880

Chapter 7
Multicellular Morphogenesis

7.1 Turing Mechanisms

The mechanisms behind generation of the visible form of advanced organisms have long been recognized among the most intriguing problems in biology. Thompson's landmark treatise *On Growth and Form* (Thompson 1917) did much to instill the need to address this problem through application of physical and mathematical methods. It was the problem of multicellular morphogenesis that motivated Turing to formulate and analyze what was to become one of the central achievements of the twentieth century science, demonstrating a mechanism for emergence of stable spatial patterns in an initially homogeneous molecular system (Turing 1952). The first situation considered was spontaneous pattern generation—a result of instability of the homogeneous equilibrium—in the distributions of two abstract diffusing reactants in a ring of cells. The emergent concentration peaks were seen as a plausible mechanism for induction of hydrozoan tentacles in the tissue around the orifice, but also for leaf primordia in the case of whorled phyllotaxis. Phyllotaxis, the developmental arrangement of the higher plants' leaves, is an example of morphogenesis that is perhaps most in evidence in the world around us. As a process whereby morphological elements (organs) are regularly and repeatedly produced to form the organism's body, it is closely related to segmentation observed in the most evolutionarily successful animal body plans. Both phyllotaxis and segmentation remain areas of intensive experimental and theoretical investigation, and each can supply excellent material with which to illustrate the presently recognized dynamical systems principles operational in multicellular morphogenesis. Before proceeding to these comparatively specialized examples, however, it is worth reviewing the Turing mechanism as originally proposed and some of this theory's successful applications.

Turing's model and the general kinetic requirements for multicellular morphogenesis that it highlights have been recently revisited by Woolley et al. (2017). Although initially formulated on a ring of cells, the model effectively deals with continuous spatial distributions characterized by densities, or local concentrations,

I. Maly, *Quantitative Elements of General Biology*,
https://doi.org/10.1007/978-3-030-79146-9_7

of molecular species (morphogens). Two morphogens are necessary for the system to exhibit the classical Turing mechanism of spatial pattern generation. Denoting the two concentrations ϕ and ψ, we can write

$$\frac{\partial \phi}{\partial t} = D_{\phi}\frac{\partial^2 \phi}{\partial x^2} + f(\phi,\psi), \qquad \frac{\partial \psi}{\partial t} = D_{\psi}\frac{\partial^2 \psi}{\partial x^2} + g(\phi,\psi), \tag{7.1}$$

where D_{ϕ} and D_{ψ} are the diffusivities of the morphogens, $x \in [0, L]$ is the spatial coordinate, and t, time. The functions f and g capture the kinetics of spatially local reactions that involve the morphogens. Zero-flux boundary conditions are frequently encountered in such models, and are compatible with a homogeneous steady state as far as the diffusional terms in Eq. (7.1) are concerned. The diffusion-driven instability that was uncovered by Turing operates around a homogeneous steady state (ϕ_0, ψ_0), in which

$$f(\phi_0,\psi_0) = 0, \qquad g(\phi_0,\psi_0) = 0.$$

For this steady state to be stable in the absence of diffusion, all eigenvalues of the Jacobian

$$J(\phi_0,\psi_0) = \left.\begin{pmatrix} \dfrac{\delta f}{\delta \phi} & \dfrac{\delta f}{\delta \psi} \\ \dfrac{\delta g}{\delta \phi} & \dfrac{\delta g}{\delta \psi} \end{pmatrix}\right|_{\phi_0,\psi_0},$$

should have a negative real part. Perturbations complying with zero-flux boundary conditions can be considered and expanded into a Fourier series,

$$\begin{pmatrix} \hat{\phi} \\ \hat{\psi} \end{pmatrix} = \sum_{m=0}^{\infty} \begin{pmatrix} a_m \\ b_m \end{pmatrix} e^{\lambda_m t} \cos(k_m x), \qquad k_m = \frac{m\pi}{L}. \tag{7.2}$$

When $\mathrm{Re}(\lambda_m) < 0$ for all m, the homogeneous steady state is stable. If, however, $\mathrm{Re}(\lambda_m) > 0$ for some $m \neq 0$, the corresponding spatial pattern (Eq. 7.2) will be amplified as the system leaves the steady state. The pattern corresponding to the largest $\mathrm{Re}(\lambda_m)$ will dominate over other components characterized by the positive values while the deviations remain small. The final solution to which the system will converge is determined by the nonlinear governing equations (Eq. 7.1). That being the general case, the dominant linear component in the Fourier decomposition (Eq. 7.2) can nevertheless serve as a reasonable ansatz for the final pattern at least in some systems. However, as pointed out already by Turing (1952), the final solution may be not only a stable pattern but a dynamic, limit-cycle spatially inhomogeneous form. In fact, the stable patterned solution may coexist with the limit cycle and even with chaotic spatially distributed forms to which the system may alternatively evolve (Aragón et al. 2012).

Dynamics of the perturbation (Eq. 7.2) near the steady state, (ϕ_0, ψ_0), is described by

$$\mathbf{0} = \begin{pmatrix} \lambda_m + D_\phi k_m^2 - f_\phi & -f_\psi \\ -g_\phi & \lambda_m + D_\psi k_m^2 - g_\phi \end{pmatrix} \begin{pmatrix} a_m \\ b_m \end{pmatrix},$$

which is the result of linearizing Eq. (7.1) and substituting Eq. (7.2). Existence of a nontrivial solution requires that

$$\lambda_m^2 + \lambda_m \left(k_m^2 \left(D_\phi + D_\psi \right) - f_\phi - g_\psi \right) + h(k_m) = 0, \tag{7.3}$$

where we have introduced

$$h(k) = D_\phi D_\psi k^4 - k^2 \left(D_\phi g_\psi + D_\psi f_\phi \right) + f_\phi g_\psi - f_\psi g_\phi. \tag{7.4}$$

From Eq. (7.3), we obtain

$$\lambda_m = \frac{f_\phi + g_\psi - k_m^2 \left(D_\phi + D_\psi \right) \pm \sqrt{\left(f_\phi + g_\psi - k_m^2 \left(D_\phi + D_\psi \right) \right)^2 - 4h(k_m)}}{2}. \tag{7.5}$$

This relationship demonstrates that in the absence of diffusion, $D_\phi = D_\psi = 0$, stability of the steady state, (ϕ_0, ψ_0), requires that

$$f_\phi + g_\psi < 0, \quad f_\phi g_\psi - f_\psi g_\phi > 0. \tag{7.6}$$

In the presence of diffusion, taking into account the conditions in Eq. (7.6), we observe that the expression before the square root in Eq. (7.5), and therefore the real part of one of the eigenvalues λ for the given m, are always negative. The other eigenvalue, however, has a positive real part when $h(k_m) < 0$.

Observing that $h(k)$ is quadratic in k^2 (Eq. 7.4), the last condition will be satisfied when

$$\begin{gathered} k_-^2 < k_m^2 < k_+^2, \\ 2D_\phi D_\psi k_\pm^2 = D_\phi g_\psi + D_\psi f_\phi \\ \pm \sqrt{\left(D_\phi g_\psi + D_\psi f_\phi \right)^2 - 4D_\phi D_\psi \left(f_\phi g_\psi - f_\psi g_\phi \right)}. \end{gathered} \tag{7.7}$$

This is possible when both the expression before the square root and the expression under it in Eq. (7.7) take positive values. That, in turn, requires

$$D_\phi g_\psi + D_\psi f_\phi > 2\sqrt{D_\phi D_\psi \left(f_\phi g_\psi - f_\psi g_\phi \right)}. \tag{7.8}$$

Summarizing, the conditions in Eqs. (7.6)–(7.8) lead to Turing instability of the homogeneous state.

Recent experimental work has uncovered diverse examples of the Turing mechanism in organ-level morphogenesis. So, experiments demonstrate a changing spatial wavelength in the digit formation zone of a mouse limb bud, depending on the dose of *Hox*-family genes (Sheth et al. 2012). Reduction of the *Hox* gene dose progressively narrows the proximal-distal zone of pattern formation in the distribution of the early chondrocyte differentiation factor Sox9, while simultaneously decreasing the spatial wavelength of the pattern along the anterior-posterior dimension. These results are recapitulated in a two-dimensional model of the forming autopod that involves regulation of a parameter analogous to g_ϕ (see above) by *Hox* gene products. Decrease of the wavelength results in polydactyly. This model also allows addressing bifurcations of the digit primordia, which represent conservation of the wavelength in a spatial domain that is broadening toward the distal edge of the bud. Although digit bifurcations are abnormal in tetrapods, they are a common pattern in homologous fin rays. Overall, the stabilization of the pentadactyl form can be seen through the prism of this study as a regulatory effect superimposed on the fundamentally flexible Turing mechanism of chondrocyte cell fate specification in vertebrate fin and limb organs.

The radiating pattern of the digital primordia has been related to the proximal-distal gradient of fibroblast growth factor (FGF), which acts as a co-modulator of the same parameter (g_ϕ) controlling the wavelength. Thus, in two dimensions, the gradient of FGF that is tuned to the broadening of the limb bud results in a diverging arrangement of the constant number of digital precursors instead of the bifurcation (Sheth et al. 2012). Subsequent work has identified Sox9 as the central node in a reaction network with diffusible products of *Wnt* and *Bmp* genes (Raspopovic et al. 2014). Vis-à-vis Sox9, Wnt acts as an inhibitor and Bmp, as an activator. This elaboration of the activator-inhibitor paradigm permitted explanation also of oligodactyly, which is observed in the experiments when the morphogenetic wavelength is increased.

Another diffusion-mediated mechanism has been identified in the formation of rugae on the mammalian palate (Economou et al. 2012). The precursor stripes in this case are associated with expression of *Shh* and appear sequentially to maintain regular spacing as the palate grows. Typically of the Turing-type dynamics, which is known from computational studies, rearrangement of the Shh pattern next to a cut made in the embryonic tissue during the development of the rugae involves appearance of branches separated by approximately 120 degrees from bent sections of the stripe that is nearest to the cut. In genetic perturbations, moreover, a similarly characteristic stripe-spot transition is seen in the forming patterns, normal rugae being replaced by a large number of "bumps" on the palate in the affected individuals. The experiments further pinpointed FGF as the diffusible activator and Shh as the diffusible inhibitor in this system.

Wnt and its inhibitor, DKK, have been implicated in the regular spacing of hair follicle buds in murine skin (Sick et al. 2006). In this system, Wnt induces both development of the bud and expression of DKK in the surrounding mesenchyme, which inhibits formation of other buds close to the given focus of Wnt expression. During the early postnatal waves of follicle induction, new follicles form in the spaces between the follicles laid down in the preceding waves. By manipulating the expression of diverse DKK family ligands, which act interchangeably as inhibitors, it is possible to effect a transition from regular spacing to formation of tight clusters of follicles that are surrounded by a ring of Wnt signaling activation and, further out, by large hairless areas. These effects correspond closely to the pattern shifts predicted by the reaction-diffusion kinetics. Overall, it is becoming clear that while multicellular pattern formation can be controlled by a variety of factors, including mechanical and nondiffusible ones, examples that involve the diffusion-driven Turing instability will also multiply with continued application of modern experimental methods.

7.2 Phyllotaxis

Phyllotaxis is the spatiotemporal pattern of initiation and growth of leaves on a stem, and also describes the similar arrangement of related organs like petals and sepals. It has long been noted that a limited number of leaf patterns are found across the species of vascular plants despite the variations in the forms of these organs and the plant habit overall. Fundamentally, organ formation on a stem can occur sequentially, or simultaneously in groups (whorls) of different number. Leaves formed one by one are oriented at a regular angle with respect to the older leaves, which in the general case produces a spiral arrangement along the shoot. If the separation angle is $\pi/2$, the sequential arrangement is termed distichous; other separation arrangements result in visible spirals, which frequently involve leaf separations that are close to the golden angle. Leaves formed in pairs, triples, and groups of four are normally separated, within the group, by the corresponding fraction of 2π, and the next group formed on the given shoot may be rotated by one-half of the dividing angle. The resulting arrangements in this case are known as dicussate (pairs that form a cross when viewed along the shoot axis), tricussate, and tetracussate. Obviously, all these patterns represent cycles of leaf formation angles in the course of the shoot development. Multiple periodicity is also encountered sufficiently often among plant species, as exemplified by orixate phyllotaxis, which is characterized by single leaves forming in a sequence of four separation angles ($\pi/2$, $\pi/4$, $\pi/2$, $-\pi/4$).

Qualitatively, the most common patterns comply with the Hofmeister rule, whereby the organ primordia are formed on the apical meristem of a shoot in the largest angular gaps between the last formed primordia and at the largest distance from the existing primordia (Hofmeister 1868). Phenomenological interaction fields can be employed to generate various types of realistic sequences of primordia formation; they are modeled as formation-inhibiting potentials and decreasing functions of the distance, on the surface of the apical meristem, that emanate from each existing primordium (Douady and Couder 1992, 1996). Mechanics of strain and buckling of the meristem epidermis explains formation of many of the patterns

(Green 1992; Green et al. 1996; Shipman and Newell 2004), as do the processes of auxin redistribution between epidermal cells (Smith et al. 2006; Jönsson et al. 2006). Discovery of the latter process (Reinhardt et al. 2000; Benková et al. 2003; Reinhardt et al. 2003) represented an unexpected spatial-kinetic twist on the abstract reaction-diffusion models of phyllotaxis (see Harrison 1993) that ascended to the Turing theory. In the recent years, a synthesis of quantitative approaches has been achieved in the study of phyllotaxis that not only identifies the mechanistic underpinnings of the most detailed quantitative treatments of this phenomenon, but can serve as a methodological example for other areas of study in multicellular morphogenesis.

Cooperation and competition between the mechanical and biochemical processes in phyllotaxis has been analyzed in detail by Newell et al. (2008) on the basis of a particularly comprehensive categorization of the foregoing approaches, to which the interested reader may be referred. The Jönsson et al. (2006) model is adopted as the starting point for the derivations. This model considers the lattice of tunica cells in the generative region of the apical meristem to be square, and the dynamics of the concentration of auxin (growth hormone) in cell i to be governed by an equation of the following type:

$$\frac{da_i}{dt} = k_d \sum_j \left(a_j - a_i\right) + k_t \sum_j \left(a_j p_{ji} - a_i p_{ij}\right) + k_s - k_o a_i. \tag{7.9}$$

Index j runs over the neighbors of cell i, and the right-hand-side terms describe the flow of auxin between the cells. The first two deal with the transport processes that occur within the layer of the tunica (epidermis): diffusion and active transport. The last two describe supply of auxin from the underlying meristem corpus and outflow into the corpus. The active transport takes place at rates proportional to the concentration of transporter protein termed PIN, p_{ij}, in the plasma membrane of cell i where it is facing cell j. The redistribution of the auxin transporter in a given cell, which is at the core of the auxin-dependent patterning of the apical meristem, can be captured by the following equations, which involve its interior concentration, p_i, in cell i:

$$\frac{dp_i}{dt} = \sum_j \left(k_{\text{int}} p_{ij} - k_{ext} a_j p_i\right),$$
$$\frac{dp_{ij}}{dt} = k_{ext} a_j p_i - k_{\text{int}} p_{ij}. \tag{7.10}$$

Constants k_{int} and k_{ext} that are introduced here describe the specific rates of internalization and externalization of PIN, the latter process occurring in proportion to the neighboring cell's auxin concentration. Indeed, since $k_{int} \gg k_o$, we can always take

$$p_{ij} = p_{tot} \frac{a_j}{\frac{k_{\text{int}}}{k_{ext}} + \sum_j a_j}, \tag{7.11}$$

where p_{tot} is the total concentration of PIN, assumed to be time-invariable and equal across the cells, and substitute this algebraic expression for the dynamics expressed by Eq. 7.10. Equations (7.9) and (7.11), therefore, constitute the auxin model with explicit cells. As demonstrated by Jönsson et al. (2006), the uniform steady-state solution is unstable when $k_t p_{tot}/k_d$ is sufficiently large—*i.e.*, when the active transport predominates sufficiently over diffusion—leading to development of a growth hormone pattern in the meristem.

Given that the characteristic scale of the emerging pattern is greater than the cell scale (as obviously should be the case in formation of organ primordia), Newell et al. (2008) formulate a continuous-space version of the auxin model dynamics for deviations g of the auxin concentration from the spatially uniform steady state:

$$g_t + D_g \nabla^4 g + H\nabla^2 g + k_o g + \kappa_1 \nabla\left(g\nabla g\right) + \kappa_2 \nabla\left(\nabla g \nabla^2 g\right) = 0,$$

$$D_g = \frac{p_{tot} k_t h^4 a_0^{\ 2}}{4\left(k_{int}/k_{ext} + a_0\right)^2}, \quad H = -h^2 k_d + \frac{p_{tot} k_t h^2 a_0^{\ 2}}{\left(k_{int}/k_{ext} + a_0\right)^2},$$

$$\kappa_1 = \frac{2 p_{tot} k_t h^2 k_{int} a_0 f_0}{k_{ext}\left(k_{int}/k_{ext} + a_0\right)^3}, \quad \kappa_2 = \frac{p_{tot} k_t h^4 k_{int} a_0^{\ 2} f_0}{2k_{ext}\left(k_{int}/k_{ext} + a_0\right)^3}. \tag{7.12}$$

Here, we have adopted an expansion as follows:

$$a_0 = k_s / k_o, \quad a\left(x,y,t\right) = a_0\left(1 + f_0 g\left(x,y,t\right)\right).$$

Notably, when the effective diffusion coefficient $-H$ becomes negative, *i.e.*,

$$k_d < \frac{p_{tot} k_t a_0^{\ 2}}{\left(k_{\mathrm{int}}/k_{ext} + a_0\right)^2},$$

the instability requires only that the perturbation be concentrated enough for the effective diffusion term to overcome the damping term that stands for the first-order outflow to the underlying meristem ($k_o g$).

To begin treatment of the meristem cap tunica mechanics along the lines of Shipman and Newell (2005), it is instructive to examine angular (dimension α) and radial (dimension r) stress-strain relationships, neglecting for the moment the Poisson ratio. We can distinguish between the total strains ε and growth-related strains G in the calculation of stresses along the two dimensions:

$$N_{\alpha\alpha} = Eh\left(\varepsilon_{\alpha\alpha} - G_{\alpha\alpha}\right), \quad N_{rr} = Eh\left(\varepsilon_{rr} - G_{rr}\right),$$

where E is Young's modulus and h, the tunica thickness. The growth strain may represent a plastic deformation or, alternatively, compression resulting from growth against a constraint. The perturbation introduced to this picture by the spatially variable, auxin-controlled growth can be captured as follows:

$$G_{\alpha\alpha} = \frac{P}{Eh} + g_{\alpha\alpha}, \quad G_{rr} = \frac{\chi P}{Eh} + g_{rr}, \quad G_{r\alpha} = g_{r\alpha},$$

where P and χ are constants. The corresponding form of the Föppl-von Kármán-Donnell stress equilibrium equation for the tunica cap (considered to be flat for simplicity) is

$$D\nabla^4 w + \chi P \frac{\partial^2}{\partial r^2} w + P \frac{1}{R^2} \frac{\partial^2}{\partial \alpha^2} w - N'_{rr} \frac{\partial^2}{\partial r^2} w$$
$$-N'_{\alpha\alpha} \frac{1}{R^2} \frac{\partial^2}{\partial \alpha^2} w + 2N'_{r\alpha} \frac{1}{R} \frac{\partial^2}{\partial r \partial \alpha} w = 0,$$
$$D = Eh^3 v^2, \quad v^2 = \frac{1}{12\left(1-\mu^2\right)}, \quad N'_{ij} = Eh\left(\varepsilon_{ij} - g_{ij}\right).$$

Here, w denotes the surface deformation and μ, the Poisson ratio. The coupling of the mechanics to biochemistry of the tunica growth regulation is completed, in the Newell et al. (2008) approach, by introducing a perturbation Airy stress function f such that

$$\frac{\partial^2 f}{\partial r^2} = N'_{rr}, \quad \frac{1}{R^2} \frac{\partial^2 f}{\partial \alpha^2} = N'_{\alpha\alpha}, \quad \frac{1}{R} \frac{\partial^2 f}{\partial r \partial \alpha} = N'_{r\alpha},$$

and adding a term to the above-derived equation for g (Eq. 7.12) that is equal to $\beta\nabla^2 f$. In this term, β represents a stress-growth coupling constant, while $\nabla^2 f$ is the trace of the stress tensor. The simple coupling form depends on an assumption that the growth tensor itself is isotropic. The coupling means that growth is linearly enhanced by local tension and suppressed by compression.

Analysis of the unified model (Newell et al. 2008) shows that the pattern of organ primordia may be driven by auxin redistribution, by mechanical buckling of the tunica, or by cooperation (resonance) between the two processes when each separately may be incapable of pattern generation. Furthermore, the patterns arising from each influence may not match spatially, yet enhance each other, resulting in a phyllotactic arrangement that is different from the primordial tunica buckling shape. Recently, the coupled mechanochemical growth model has been elaborated by Rueda-Contreras et al. (2018). Their three-dimensional simulations reproduced a number of frequently encountered and complex phyllotactic patterns, including the tricussate and tetracussate sequences of primordia (Fig. 7.1). Further work on the analysis of the integrated mechanochemical mechanism has been done by Julien et al. (2019).

The explanatory power of the combined mechanochemical model can be exploited by reducing its mathematical form to that which had been utilized

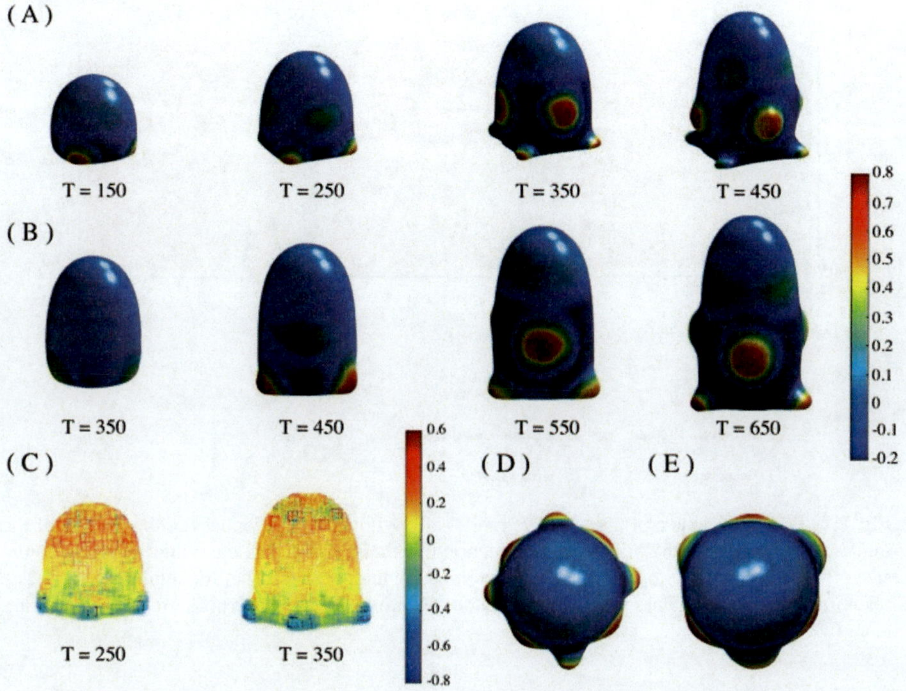

Fig. 7.1 Development of phyllotactic patterns in a coupled mechanochemical growth model. (**a**) Tetracussate phyllotaxis (**b**) Tricussate phyllotaxis (**c**) Curvature in the tetracussate model (**d**) Top view of the tetracussate model (**e**) Top view of the tricussate model. Simulated times are indicated. Reproduced with permission from Rueda-Contreras et al. (2018)

successfully in phenomenological generative models of phyllotactic sequences. The quantitative connection between the mechanochemical model output and the input parameters of Douady-Couder style models has been pointed out by Mirabet et al. (2012). Indeed, whether they are determined by the mechanical or by the biochemical submodel, the patterns that arise on the generational ring in the apical tunic have the form

$$A\cos(2\pi x/d) + B\cos(4\pi x/d).$$

The factors that are found here can be expressed, in particular, via the parameters of Eq. (7.12):

$$A = \frac{2D_g\sqrt{(H^2 - 4k_o D_g)(2H^2 + k_o D_g)}}{H\sqrt{(H\kappa_2 - 2\kappa_1 D_g)(H\kappa_2 + \kappa_1 D_g)}}, \quad B = -\frac{D_g(H^2 - 4k_o D_g)}{H(H\kappa_2 + \kappa_1 D_g)},$$

$$d = 2\pi\sqrt{2D_g/H}.$$

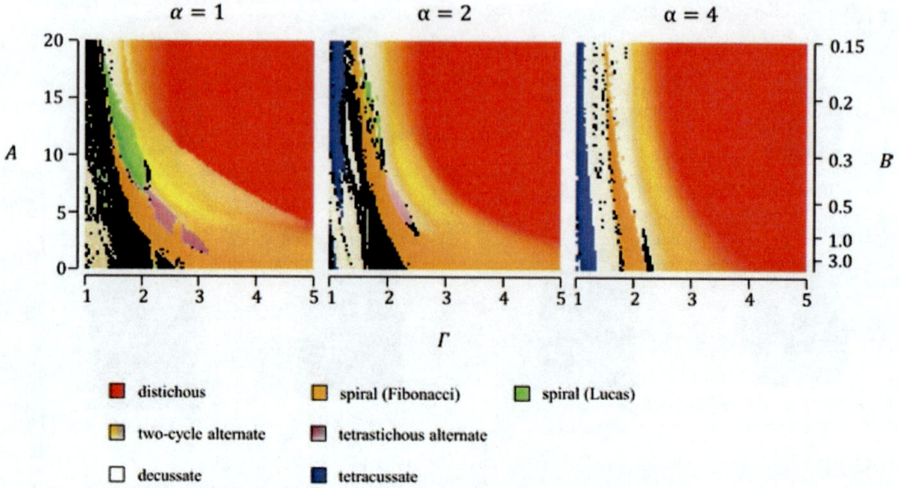

Fig. 7.2 Parameter space of a generative model of phyllotaxis. α, steepness of the decline of the inhibitory field around the primordium induction threshold, Γ, ratio of the inhibition range to the size of the apical meristem, A and B, parameters of the temporal decay of the inhibitory field. The resulting phyllotactic classes are color-coded. Reproduced with permission from Yonekura et al. (2019)

In terms of phenomenological inhibitory field descriptors that define the interaction between existent and forming primordia, the inhibition range is d, the threshold of inhibition is approximately the ratio of the smallest (a_0–A–B) to the largest (a_0–A–B) auxin level, and the steepness of the field decline with distance is B/A. In principle, a similar treatment is possible with the use of the mechanical parameters in the instances where they govern the pattern, and for the parameters of the mechano-chemical resonance, when the latter is defining.

The derivation of the inhibitory-field model parameters from the coupled growth formalism lends support to a new, mechanistic interpretation of generative models of phyllotaxis. Continued development of this type of models has met with considerable success. Most recently, Yonekura et al. (2019) performed an extensive analysis of the parameter space of a generative model featuring a time-dependent decline in the inhibitory strength in addition to the usually modeled distance-dependent decline, and offered a mechanistic interpretation of such a property of the primordial inhibitory field. This extension permitted an explanation of the four-cycle orixate phyllotaxis, and additionally resulted in characterization of kinetic relationships between a large number of phyllotactic patterns in the extended parameter space of this sophisticated generative model (Fig. 7.2). Overall, the ongoing efforts in the field of phyllotaxis present an example of successful integration of biomechanical, biochemical, and more abstracted dynamic systems approaches to tackle the complexity of multicellular morphogenesis.

7.3 Segmentation

Most of the readily observable animals are segmented. Indeed, while Mollusca do not have segments, Annelida, Insecta, and Vertebrata do, as emphasized even by the common names of these clades. The modern classification (see Ten Tusscher 2013) shows that segmented groups occur in separate, diverse larger clades of Bilateria: Chordata among other, non-segmented Deuterostomia, such as Echinodermata; Arthropoda among non-segmented Ecdysozoa such as Priapulida; Annelida alongside the already mentioned Mollusca among other Lophotrochozoa. Although the classification of clades as segmented and non-segmented, and their arrangement into higher-level groups is not entirely settled, the noted aspect of the segmented animals' distribution on the tree of life appears able to withstand this uncertainty. One obvious interpretation of this picture is that segmentation can arise readily, at least in the bilaterian developmental plan, and when it does, it enables adaptive radiation to a degree rarely matched by non-segmented body organization. This view may be biased toward niches populated by us, segmented animals ourselves, but it may also be general as far as advanced multicellular life forms are concerned. In either case, the considerable attention devoted to the mechanisms of segmentation in recent research has led to a level of understanding that presents interest as a paradigm for unraveling multicellular development from the standpoint of dynamic systems.

The notion that segmentation is an expected rather than accidental type of developmental organization is supported by simulations of evolutionary trajectories in models of developmental regulatory networks (see Ten Tusscher 2013). The results of this line of investigation also suggest that multiple pathways to the emergence of segmentation may indeed exist. Segmentation patterns evolve in the simulations if a symmetry-breaking signal is present, for example, in the form of a spatial perturbation in gene expression, an embryo-scale gradient of a morphogenetic molecule—or a propagating wavefront, which is a condition of special relevance to vertebrate segmentation that will be considered in detail later in this section. An interesting theoretical result of the evolutionary simulations is that even if the set of genes involved in rostro-caudal differentiation is allowed to expand indefinitely, segmentation nonetheless arises reproducibly as a function of a very limited number of interacting gene products, with the architecture of the mini-network, rather than finely tailored functionality of the genes, being responsible for the emergence of the pattern. In fact, only a few modes of segmented development arise, largely irrespectively of the details assumed about the simulated evolutionary process. This result is believed to reflect fundamental biological constraints that require that any arising mechanism be both evolutionarily accessible and robust with respect to further alterations.

Although the modes of segmentation that are reproducibly evolving in simulations can be broken down along different lines, for example with regard to the architecture of the governing gene interaction network, the distinction between sequential and simultaneous segmentation mechanisms that is recapitulated in the evolutionary-developmental simulation studies is the one that parallels the classification known already from classical embryology (see Ten Tusscher 2013; Pais-de-Azevedo et al. 2018). The sequential modes correspond to specific implementations of the clock-and-wavefront mechanism, which was originally proposed as a hypothesis (Cooke and Zeeman 1976) and later substantiated and elaborated in experiments on vertebrate embryos (see Pais-de-Azevedo et al. 2018). In this type of segmentation, the segments—in the vertebrate case, somites—are laid down one by one as propagation of a differentiation front in the caudal direction causes spatial readout and fixation of a temporal pattern that exists autonomously in the form of biochemical and genetic oscillations in the caudal bud of the embryo that elongates at its posterior end. Simultaneous segmentation, by contrast, arises as hierarchical regulation among a set of gene classes, whose number grows with the number of simultaneously laid down segments. This mechanism emerging in the evolutionary simulations recapitulates, in its main system-level features, the one established by the groundbreaking experiments on fruit fly embryos (Nüsslein-Volhard and Wieschaus 1980; Rivera-Pomar and Jäckle 1996). In the simulations, as well as among the studied developmental types of insects, the simultaneous segmentation may encompass the entire embryo, or it may result in a simultaneous deposition of a certain number of rostral segments, whereas the more caudal segments will form sequentially. In the latter case, characteristically, the number of the simultaneously laid down segments may be conserved and the number of the remaining segments, more variable among related species (*i.e.*, of their larvae). To emphasize the apparent universality of this principle, it may be noted that a small number of the most rostral somites in vertebrates are also laid down simultaneously. Overall, the sequential mechanisms appear primordial, and simultaneous ones, derived and specialized (Ten Tusscher 2013).

The basic kinetics of the clock part of the clock-and-wavefront model has found a solid foundation in both experiments and mathematical analyses. The molecular implementation of the oscillator in the cells of vertebrate presomitic mesoderm at the caudal bud varies somewhat between the studied species, but typically involves Wnt, Notch, and FGF (fibroblast growth factor) genes and their products (see Pais-de-Azevedo et al. 2018). The kinetic nature of these oscillations resembles and incorporates elements of the paradigmatic oscillatory signaling and gene expression processes that we have considered in the chapter on general biological regulation. Importantly, while the oscillations are intrinsic to the cells of the presomitic mesoderm, their synchronization among the neighboring cells, and fine-tuning of the oscillation period, depend on coordination between cells that is mediated by secreted Notch ligands.

At the same time, the nature of the wavefront part of the Cooke-Zeeman model is not fully settled by molecular-biological experiments (Pais-de-Azevedo et al. 2018). The wavefront of cell differentiation propagates posteriorly in coordination with the embryo's growth at the caudal end and appears to "read out" the oscillations in the presomitic mesoderm, resulting in commitment of the cells to developmental identity that defines their place in the somites. The wave of differentiation has long been known to be correlated with gradients of Wnt and FGF, which emanate from the caudal bud, and a countergradient of retinoic acid emanating from formed somites. How exactly (or whether) these gradients result in "freezing" of the clock waves that is observed in the cell differentiation zone, remains difficult to deduce from experiments in purely descriptive terms. Mathematical analysis conducted to date, however, explains the translation of oscillations into a stable spatial pattern in the growing domain as an emergent feature of the spatially distributed oscillations themselves.

Murray et al. (2011) laid out a clear demonstration of a stable differentiation wavefront in a spatially distributed system of oscillators with kinetic properties resembling the Wnt-Notch-FGF system in vertebrate embryos. Given a chain of coupled oscillators (which can be thought of as stretching anterio-posteriorly in an embryo), dynamics of substance concentrations u and v in oscillator k can be assumed, intrinsically, to follow the law

$$\dot{u}_k = \omega v_k + P u_k\left(1 - u_k^2 - v_k^2\right),$$
$$\dot{v}_k = -\omega u_k + P v_k\left(1 - u_k^2 - v_k^2\right). \tag{7.13}$$

Here, ω is the frequency and P, the strength of attraction of the limit cycle. Transforming the coordinates,

$$u_k = r_k \cos\theta_k, \qquad v_k = r_k \sin\theta_k, \tag{7.14}$$

we can replace Eq. (7.13) with

$$\dot{r}_k = \mathrm{Pr}_k\left(1 - r_k\right), \qquad \dot{\theta}_k = \omega. \tag{7.15}$$

The dynamics of Eq. (7.15) relax to

$$r_k = 1, \qquad \dot{\theta}_k = \omega. \tag{7.16}$$

Now, let us denote the linear and quadratic coupling strength between the neighboring oscillators as $\tilde{A}$ and $\tilde{B}$. Eq. (7.13) can then be expanded as follows:

$$\dot{u}_k = \omega v_k + P u_k\left(1 - u_k^2 - v_k^2\right) + \tilde{A}\left(u_{k-1} - 2u_k + u_{k+1}\right)$$
$$-\tilde{B} v_k\left(\left(u_{k+1} - u_k\right)^2 + \left(v_{k-1} - v_k\right)^2\right),$$
$$\dot{v}_k = -\omega u_k + P v_k\left(1 - u_k^2 - v_k^2\right) + \tilde{A}\left(v_{k-1} - 2v_k + v_{k+1}\right)$$
$$-\tilde{B} u_k\left(\left(u_{k+1} - u_k\right)^2 + \left(v_{k-1} - v_k\right)^2\right). \tag{7.17}$$

Applying the coordinate transformation (Eq. 7.14) now to Eq. (7.17), we obtain, in the limit of $P \gg A$,

$$\frac{d\theta_k}{dt} = \omega + \tilde{A}\left(\theta_{k-1} - 2\theta_k + \theta_{k+1}\right) - \frac{\tilde{B}}{2}\left(\left(\theta_k - \theta_{k-1}\right)^2 + \left(\theta_k - \theta_{k+1}\right)^2\right),$$
$$r_k \sim 1. \tag{7.18}$$

To transition to a continuous description of the system along the anterio-posterior axis x, let us introduce d, the cell size, and consider

$$\Delta k = d\Delta x, \qquad A = \tilde{A}\Delta k^2, \qquad B = \tilde{B}\Delta k^2. \tag{7.19}$$

Then, under certain realistic conditions specified by Murray et al. (2011), we can write the continuous form of Eq. (7.18) as follows:

$$\frac{\partial \theta}{\partial t} = \omega + A\frac{\partial^2 \theta}{\partial x^2} - B\left(\frac{\partial \theta}{\partial x}\right)^2. \tag{7.20}$$

The synchrony of oscillations in the posterior presomitic mesoderm and limit-zero oscillation rate at the remote anterior end of the embryo can be captured by the following boundary conditions:

$$\left.\frac{\partial \theta}{\partial x}\right|_{x\to\infty} = 0, \qquad \left.\frac{\partial \theta}{\partial x}\right|_{x\to-\infty} = \sqrt{\frac{\omega}{B}}. \tag{7.21}$$

The latter value is the steady-state solution to Eq. (7.20).

Central to the concept of spontaneous fixation of the temporal oscillations into a spatial pattern is the description of the introduced system dynamics that focuses on phase gradient

$$\Psi(x,t) = \frac{\partial \theta}{\partial x}. \tag{7.22}$$

The phase gradient evolves according to Burger's equation,

$$\frac{\partial \Psi}{\partial t} + 2B\Psi\frac{\partial \Psi}{\partial x} = A\frac{\partial^2 \Psi}{\partial t^2}, \tag{7.23}$$

known to have a traveling-wave solution. Denoting the coordinate of the traveling wave $y = x–Ut$, we obtain that $U = \sqrt{\omega B}$ and have

$$\theta(x,t) \sim \omega t, \qquad \theta(x,t) \sim \sqrt{\frac{\omega}{B}}x \tag{7.24}$$

ahead of the wave and behind the wave, correspondingly. The spatial wavelength of the somite pattern behind the wave, therefore, will be

$$S = 2\pi\sqrt{\frac{B}{\omega}}. \tag{7.25}$$

The phase gradient at y, when the traveling waveform becomes established, is

$$\Psi(y) = \frac{\sqrt{\omega / B}}{1 + e^{\sqrt{\omega B}\, y / A}}. \tag{7.26}$$

Using the last result (Eq. 7.26) as an initial condition, solutions to Eq. (7.20) with the already specified boundary conditions (Eq. 7.21) can be obtained that represent idealized dynamics of somitogenesis along the anterio-porterior axis of a vertebrate embryo (Murray et al. 2011). By plotting $\sin\theta$ as a function of x and t (Fig. 7.3a–d), we can observe progressive transformation of largely synchronous temporal oscillations at the posterior end into elements of a stable spatial pattern whose domain expands from the anterior. The strongly nonlinear increase of the phase gradient (Fig. 7.3e) underlies, according to the coupled oscillators formalism, convergence of the spatial wavelength to a constant value (Fig. 7.3f), which is the hallmark of early vertebrate development.

Detailed simulations of somitogenesis, including spatial cell dynamics, were carried out by Hester et al. (2011). More recently, it proved possible to evolve *in silico* a simple gene interaction network that exhibits an emergent differentiation wavefront (Cotterell et al. 2015). Analysis of this network's spatial kinetics, which

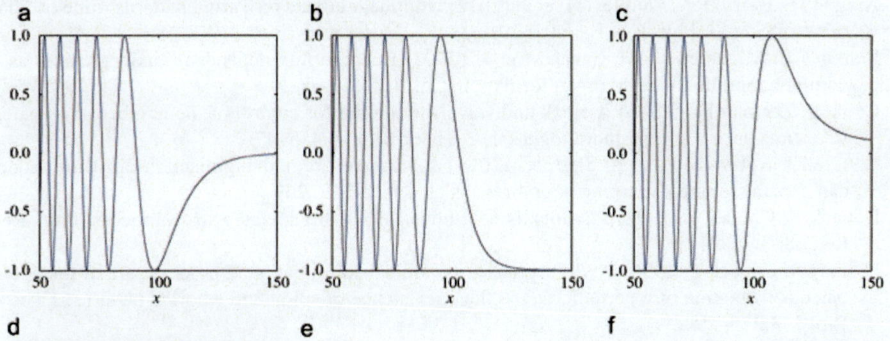

Fig. 7.3 Differentiation wavefront propagation in a coupled oscillators model of somitogenesis. (**a-d**) oscillator phase in time and space, (**e**) phase gradient, (**f**) spatial wavelength. Propagation of the differentiation front is shown with arrows. The anterior of the embryo is to the left. Reproduced from Murray et al. (2011). Reprinted from Journal of Theoretical Biology, Vol. 283, Philip J. Murray, Philip K. Maini, Ruth E. Baker, The clock and wavefront model revisited, 227–238, copyright 2011, with permission from Elsevier

includes posterior oscillations and their spontaneous fixation in the form of a posteriorly propagating differentiation wave, led to a molecular interpretation of the differentiation front as arising from inhibition of the oscillatory dynamics by the formed anterior pattern. This model appears fully compatible with the more abstract mathematical analysis of an oscillatory medium as laid out above. At the same time, it eliminates the residual conceptual difficulty in applying the coupled oscillators framework to somite differentiation treated as an emergent developmental phenomenon. This difficulty was connected with the Murray et al. assumption of a constant phase gradient at the anterior end of the embryo, and the absence of a need for such an assumption in the Cotterell model demonstrates the emergent properties of the system more unambiguously. Moreover, the Cotterell mechanism can, with parameter changes, exhibit simultaneous segmentation as well. The capabilities of the modern dynamic systems approach to somite differentiation, therefore, hold out a promise of quantitative integration of our understanding of the two fundamental types of segmentation processes.

References

Aragón JL, Barrio RA, Woolley TE et al (2012) Nonlinear effects on Turing patterns: time oscillations and chaos. Phys Rev E 86:026201

Benková E, Michniewicz M, Sauer M et al (2003) Local, efflux-dependent auxin gradient as a common module for plant organ formation. Cell 115:591–602

Cooke J, Zeeman EC (1976) A clock and wavefront model for control of the number of repeated structures during animal morphogenesis. J Theor Biol 58:455–476

Cotterell J, Robert-Moreno A, Sharpe J (2015) A local, self-organizing reaction-diffusion model can explain somite patterning in embryos. Cell Syst 1:257–269

Douady S, Couder Y (1992) Phyllotaxis as a physical self-organized growth process. Phys Rev Lett 68:2098–2101

Douady S, Couder Y (1996) Phyllotaxis as a dynamical self organizing process part II: the spontaneous formation of a periodicity and the coexistence of spiral and whorled patterns. J Theor Biol 178:275–294

Economou AD, Ohazama A, Porntaveetus T et al (2012) Periodic stripe formation by a Turing mechanism operating at growth zones in the mammalian palate. Nat Genet 44:348–351

Green PB (1992) Pattern formation in shoots: a likely role for minimum energy configurations of the tunica. Int J Plant Sci 153:S59–S75

Green PB, Steele CS, Rennich SC (1996) Phyllotactic patterns: a biophysical mechanism for their origin. Ann Botany 77:515–527

Harrison LG (1993) Kinetic theory of living pattern. Cambridge University Press, Cambridge

Hester SD, Belmonte JM, Gens JS et al (2011) A multi-cell, multi-scale model of vertebrate segmentation and somite formation. PLoS Comput Biol 7:e1002155

Hofmeister W (1868) Allgemeine Morphologie der Gewächse. Engelmann, Leipzig

Jönsson H, Heisler M, Shapiro B et al (2006) An auxin-driven polarized transport model for phyllotaxis. Proc Natl Acad Sci U S A 103:1633–1638

Julien JD, Pumir A, Boudaoud A (2019) Strain- or stress-sensing in mechanochemical patterning by the phytohormone auxin. Bull Math Biol 81:3342–3361

Mirabet V, Besnard F, Vernoux T et al (2012) Noise and robustness in phyllotaxis. PLOS Comp Biol 8:e1002389

Murray PJ, Maini PK, Baker RE (2011) The clock and wavefront model revisited. J Theor Biol 283:227–238
Newell AC, Shipman PD, Sun Z (2008) Phyllotaxis: cooperation and competition between mechanical and biochemical processes. J Theor Biol 251:421–439
Nüsslein-Volhard C, Wieschaus E (1980) Mutations affecting segment number and polarity in *Drosophila*. Nature 287:795–801
Pais-de-Azevedo T, Magno R, Duarte I, Palmeirim I (2018) Recent advances in understanding vertebrate segmentation. F1000 Res 7:97
Raspopovic J, Marcon L, Russo L et al (2014) Digit patterning is controlled by a bmp-Sox9-Wnt Turing network modulated by morphogen gradients. Science 345:566–570
Reinhardt D, Mandel T, Kuhlemeier C (2000) Auxin regulates the initiation and radial position of plant lateral organs. Plant Cell 12:507–518
Reinhardt D, Pesce E, Stieger P et al (2003) Regulation of phyllotaxis by polar auxin transport. Nature 426:255–260
Rivera-Pomar R, Jäckle H (1996) From gradients to stripes in *Drosophila* embryogenesis: filling in the gaps. Trends Genet 12:478–483
Rueda-Contreras MD, Romero-Arias JR, Aragón JL et al (2018) Curvature-driven spatial patterns in growing 3D domains: a mechanochemical model for phyllotaxis. PLoS One 13:e0201746
Sheth R, Marcon L, Bastida MF et al (2012) *Hox* genes regulate digit patterning by controlling the wavelength of a Turing-type mechanism. Science 338:1476–1480
Shipman PD, Newell AC (2004) Phyllotactic patterns on plants. Phys Rev Lett 92:168102
Shipman PD, Newell AC (2005) Polygonal planforms and phyllotaxis on plants. J Theor Biol 236:154–197
Sick S, Reinker S, Timmer J et al (2006) WNT and DKK determine hair follicle spacing through a reaction-diffusion mechanism. Science 314:1447–1750
Smith RS, Guyomarc'h S, Mandel T et al (2006) A plausible model of phyllotaxis. Proc Natl Acad Sci U S A 103:1301–1306
Ten Tusscher KH (2013) Mechanisms and constraints shaping the evolution of body plan segmentation. Eur Phys J E 36:54
Thompson DW (1917) On growth and form. Cambridge University Press, Cambridge
Turing AM (1952) The chemical basis of morphogenesis. Philos Trans R Soc Lond B 237:37–72
Woolley TE, Baker RE, Maini PK (2017) Turing's theory of morphogenesis: where we started, where we are, and where we want to go. In: Cooper SB, Soskova MI (eds) The incomputable. Springer, Cham, pp 219–235
Yonekura T, Iwamoto A, Fujita H et al (2019) Mathematical model studies of the comprehensive generation of major and minor phyllotactic patterns in plants with a predominant focus on orixate phyllotaxis. PLoS Comput Biol 15:e1007044

Chapter 8
Interspecies Dynamics

Check for updates

8.1 Coexistence and Exclusion

Dynamics arising from species interactions encompass biomass changes associated with each species, as well as influences on the evolutionary trajectory of the species, that are attributable to its ecological relationship with other species, such as the predator-prey relationship, mutualism, or competition. These dynamics, therefore, have temporal, spatial, and evolutionary aspects, and may lead to self-organized effects in each of the three domains. The biomass dynamics, obviously, is a multi-species generalization of the elementary kinetic laws that we considered in Chap. 2, while the ecological dynamics in the genotype space, as we shall see, are derivative from the elementary formalizations for speciation (see Chap. 3). Interspecies dynamics are often considered to be a higher-level manifestation of intra-organism and cellular processes. However, it is equally important to note the constraining effect of the conditions arising from interspecies dynamics on the cellular and physiological processes that may take place in the realm of the given ecosystem. Interspecies dynamics, in addition, have played a special role in the development of not only quantitative approaches to biology, but also theory of dynamical systems. With this in mind, we will open this chapter by considering regimes of coexistence and mutual exclusion in species interaction, which illustrate the general themes of stability and instability in multi-species systems. This is one of the most elaborated areas in quantitative biology, and for two interacting species, a complete theory of possible non-spatial dynamic regimes exists (reviewed in Murray 2002).

One of the most important and widespread types of species relationships is mutualism, whereby the carrying capacity of the environment is effectively increased for each of the involved species as a result of the interaction. Strong mutualism may be a requirement for survival of the populations. The dynamics of mutualistic relationships has been analyzed, in particular, by May (1975) and Whittaker (1975). Denoting the environment carrying capacities for two species in isolation K_1, K_2, we can write

I. Maly, *Quantitative Elements of General Biology*,
https://doi.org/10.1007/978-3-030-79146-9_8

$$\frac{dN_1}{dt} = r_1 N_1 \left(1 - \frac{N_1}{K_1} + b_{12}\frac{N_2}{K_1}\right), \quad \frac{dN_2}{dt} = r_2 N_2 \left(1 - \frac{N_2}{K_2} + b_{21}\frac{N_1}{K_2}\right), \tag{8.1}$$

where N is the population size, r reference reproduction rate, and b interaction coefficients (see Murray 2002). Equation (8.1) nondimensionalizes to

$$\frac{du_1}{d\tau} = u_1\left(1 - u_1 + a_{12}u_2\right), \quad \frac{du_1}{d\tau} = \rho u_2\left(1 - u_2 + a_{21}u_1\right), \tag{8.2}$$

where we have used

$$\begin{gathered} u_1 = \frac{N_1}{K_1}, \quad u_2 = \frac{N_2}{K_2}, \quad \tau = r_1 t, \quad \rho = \frac{r_2}{r_1}, \\ a_{12} = b_{12}\frac{K_2}{K_1}, \quad a_{21} = b_{21}\frac{K_1}{K_2}. \end{gathered} \tag{8.3}$$

It can be seen from Eq. (8.2) that our sample mutualistic system has steady states corresponding to extinction of one or both interacting populations: (0, 0), (1, 0), (0, 1). One more steady state is found in the biologically meaningful positive quadrant of the phase space (u_1, u_2) when $a_{12}a_{21} < 1$, *i.e.*, when the mutually beneficial interactions are not simultaneously too strong:

$$\left(\frac{1 + a_{12}}{1 - a_{12}a_{21}}, \frac{1 + a_{21}}{1 - a_{12}a_{21}}\right). \tag{8.4}$$

Applying standard phase plane analysis (see Appendix), we can observe that the first three equilibria are unstable. If the fourth equilibrium (Eq. 8.4) is found outside the positive quadrant, the positive system trajectories all lead to infinity, indicating unbounded population growth. In this case, dynamics encapsulated in Eq. (8.1) does not capture ecological factors limiting proliferation of the two species. When, however, the fourth steady state solution (Eq. 8.4) is positive, the phase plane analysis shows it to be stable. The kinetic conditions of the species interaction then support establishment and maintenance of species coexistence.

Competition between species, for example for a limited resource, can be analyzed similarly, by changing the sign of the interaction:

$$\frac{dN_1}{dt} = r_1 N_1 \left(1 - \frac{N_1}{K_1} - b_{12}\frac{N_2}{K_1}\right), \quad \frac{dN_2}{dt} = r_2 N_2 \left(1 - \frac{N_2}{K_2} - b_{21}\frac{N_1}{K_2}\right). \tag{8.5}$$

The competitive system (Eq. 8.5) accepts the same nondimensionalization as above (Eq. 8.3), leading to

$$\frac{du_1}{d\tau} = u_1\left(1 - u_1 - a_{12}u_2\right) = f_1\left(u_1, u_2\right),$$
$$\frac{du_1}{d\tau} = \rho u_2\left(1 - u_2 - a_{21}u_1\right) = f_2\left(u_1, u_2\right),$$

which differs only in the sign of the interaction terms. It, therefore, has the same three steady states characterized by extinction of one or both populations. The coexistence solution, however, has a different sign in the numerator (cf. Eq. 8.4):

$$\left(u_1^*, u_2^*\right) = \left(\frac{1 - a_{12}}{1 - a_{12}a_{21}}, \frac{1 - a_{21}}{1 - a_{12}a_{21}}\right). \tag{8.6}$$

As a result, while $a_{12}a_{21} \neq 1$ is again a requirement for the biological relevance of the last solution—easily met by any realistic parameter set—additional dynamic possibilities arise in the competitive case as the interaction parameters, a_{12} and a_{21}, are varied.

Equilibrium (0, 0) is always unstable, because (see Appendix) the corresponding eigenvalues of the interaction matrix,

$$A = \begin{pmatrix} \frac{\delta f_1}{\delta u_1} & \frac{\delta f_1}{\delta u_2} \\ \frac{\delta f_2}{\delta u_1} & \frac{\delta f_2}{\delta u_2} \end{pmatrix},$$

are always positive: $\lambda_1 = 1$, $\lambda_2 = \rho$. The competing species, therefore, do not become extinct at the same time. Equilibrium point (1, 0), which corresponds to extinction of the second species, is stable if $a_{21} > 1$, and unstable if $a_{21} < 1$. Conversely, steady state (0, 1) is stable when $a_{12} > 1$ and unstable when $a_{12} < 1$. At the equilibrium point defined in Eq. (8.6), we have

$$\lambda_{1,2} = \frac{c \pm \sqrt{c^2 - 4\rho\left(1 - a_{12}a_{21}\right)\left(a_{12} - 1\right)\left(a_{21} - 1\right)}}{2\left(1 - a_{12}a_{21}\right)},$$
$$c = a_{12} - 1 + \rho\left(a_{21} - 1\right).$$

Overall, we can see that when both a_{12} and a_{21} are smaller than 1, equilibrium $\left(u_1^*, u_2^*\right)$ lies in the positive quadrant and is stable, while the other equilibria are all unstable. The outcome of competition under these conditions is coexistence of the species. When one of the dimensionless interaction coefficients is greater and the other smaller than 1, the corresponding equilibrium describing extinction of one of the species is stable. At the same time, the alternative extinction equilibrium is unstable and the coexistence equilibrium does not exist in the biologically meaningful positive quadrant of the system phase plane. These situations describe complete displacement of one species by the other in the competitive niche.

Finally, if both a_{12} and a_{21} are greater than 1, $\left(u_1^*, u_2^*\right)$ again lies in the positive quadrant. However, it is now unstable and marks the boundary in the system's phase space, on the two sides of which the system trajectories converge to the alternative species extinction equilibria. When the dimensionless competition coefficients are large, therefore, the ecosystem is bistable, supporting one or the other competitive exclusion outcome, depending on which species is numerically favored at the outset. Such a system can also be forced, for example by a migration event, to the complementary attraction domain on the phase plane, and the outcome of the competition changed.

8.2 Temporal Patterns

Temporal behavior captured by the simple ecological-kinetic model of Lotka and Volterra (see Murray 2002) has a special place in the twentieth century science, having contributed to the recognition of autonomous, undamped oscillations as an elementary self-organization process. The Volterra model for species interaction through predation implies an intrinsically unbounded growth of the prey, a first-order intrinsic decay of the predator biomass, and a biomass transfer from the prey to the predator that is linear in the biomass of each species:

$$\frac{dN}{dt} = N\left(a - bP\right), \qquad \frac{dP}{dt} = P\left(cN - d\right).$$

Here, we have denoted the biomass of the prey as N, that of the predator as P, and assumed that each of the constants a, b, c, d is positive. These equations are known as Lotka-Volterra equations because their form corresponds to Lotka's kinetic model that preceded Volterra's ecological investigation (Lotka 1920; Lotka 1925; Volterra 1926). Nondimensionalization leads to

$$\frac{du}{d\tau} = u\left(1 - v\right), \qquad \frac{dv}{d\tau} = \alpha v\left(u - 1\right), \qquad (8.7)$$
$$u = cN / d, \qquad v = bP / a, \qquad \tau = at, \qquad \alpha = d / a.$$

By considering

$$\frac{dv}{du} = \alpha \frac{v\left(u - 1\right)}{u\left(1 - v\right)},$$

we can see that the singular points lie at $u = v = 0$ and $u = v = 1$. The phase trajectories can be calculated exactly:

$$\alpha u + v - \ln u^{\alpha} v = H.$$

H takes its minimum value, $H_{min} = 1 + \alpha$, at the second singular point. For each $H > H_{min}$, the system ascribes a closed loop trajectory around the said singular point, whose characteristic radius grows with the selected H. In the space of u, v, the trajectories appear as smoothed right triangles with two sides extending in parallel to the axes. Perturbations shift the system trajectory to a loop with a new H, instead of vanishing to allow the system to return to the original trajectory. Thus, a Lotka-Volterra predator-prey system that has settled on the cyclical behavior exhibits a type of neutral stability.

The behavior of the system can be examined in more detail by, first, linearizing about (0, 0). Doing so, we can obtain from Eq. (8.7) the following equation for perturbations x, y around this singular point:

$$\begin{pmatrix} \dfrac{dx}{d\tau} \\ \dfrac{dy}{d\tau} \end{pmatrix} \approx \begin{pmatrix} 1 & 0 \\ 0 & -\alpha \end{pmatrix} \begin{pmatrix} x \\ y \end{pmatrix} = A \begin{pmatrix} x \\ y \end{pmatrix} \tag{8.8}$$

A here denotes this ecological system's community matrix, whose elements quantify the strength and sign of the species interactions, as far as these interactions' influences on the species biomass are concerned. The dynamics encapsulated in Eq. (8.8) leads to

$$\begin{pmatrix} x \\ y \end{pmatrix} = \mathbf{B} e^{\lambda \tau},$$

with $\mathbf{B}$ being a constant column vector. The eigenvalues λ arise from the characteristic polynomial of matrix A:

$$|A - \lambda I| = \begin{vmatrix} 1-\lambda & 0 \\ 0 & -\alpha - \lambda \end{vmatrix} = 0 \quad \Rightarrow \quad \lambda_1 = 1, \quad \lambda_2 = -\alpha.$$

Since one of these is greater than zero, (0, 0) is an unstable equilibrium—specifically, a saddle point, since the second eigenvalue is negative.

Proceeding similarly for singular point (1, 1), we obtain $\lambda_1, \lambda_2 = \pm i\sqrt{\alpha}$, and, therefore,

$$\begin{pmatrix} x \\ y \end{pmatrix} = \mathbf{l} e^{i\sqrt{\alpha}\tau} + \mathbf{m} e^{-i\sqrt{\alpha}\tau}, \tag{8.9}$$

$\mathbf{l}$ and $\mathbf{m}$ being the corresponding eigenvectors. The purely imaginary eigenvalues signify a center type of singularity at (1, 1), which is neutrally stable. The cyclical solutions in the neighborhood of this point, which are implied by Eq. (8.9), have a period of $2\pi / \sqrt{\alpha}$. Although certain details of the Lotka-Volterra model for predation-mediated species interaction are now considered unrealistic, and numerous improvements to it have been formulated that comport with a wider range of empirical data (see Murray 2002), the original model laid out here has a special place in the development of modern science and remains a frequently studied exemplar of temporal pattern formation in interacting biological populations.

Interspecies dynamics, in general, should be considered in conjunction with coevolution of the species, which is accompanied by evolution of their ecological relationships. The most interesting temporal pattern identified in the recent work on this subject is apparently non-periodic (or very long-period) transitions between distinct dynamic states. The transitions correlate qualitatively with punctuated equilibrium that is described in evolutionary paleontological studies (see references in Chap. 3). These findings extend the observation of saltatory dynamics in the evolving system's behavior to the ecological level.

Species extinction punctuating periods of diversifying speciation has been addressed in analyses based on a representation of ecological relationships directly on the species level (Kauffman and Johnsen 1991; Bak and Sneppen 1993; Solé and Manrubia 1996). The type of dynamics that leads to extinction, and, generally, to "coevolutionary avalanches" in these analyses was termed self-organized criticality (see a recent re-analysis in Murase and Rikvold 2018). This approach has been further developed to make evolution in the genome space explicit (Christensen et al. 2002; Jensen 2004). Instead of proceeding from the assumption that discrete species with defined ecological characteristics exist, the genetic models demonstrate how such ecologically defined genotype clusters emerge and co-evolve. The genetic modeling method was termed "tangled nature," in an allusion to Darwin's word choice in the closing paragraph of his opus.

The tangled-nature model is based on the same description of genomes that we adopted in Chap. 3, following Higgs and Derrida (1992). To wit, a vector,

$$\mathbf{S}^{\alpha} = \left(S_1^{\alpha}, S_2^{\alpha}, \ldots, S_L^{\alpha}\right),$$

is put in correspondence to each individual (α) in the genotype space. The individuals can be α, β, ... = 1, 2, ..., $N(t)$, up to the number of individuals at time t. The elements of the genetic vector take the values of plus or minus one, to denote the alternative alleles. Furthermore, we can refer to positions (occupied or unoccupied) in the genome space using Roman superscripts:

$$\mathbf{S}^{a},\ \mathbf{S}^{b},\ \ldots,\qquad a,b,\ldots = 1,2,\ldots,2^{L}.$$

The individual's reproductive ability is assumed to be

$$H\left(\mathbf{S}^{\alpha},t\right) = \frac{1}{N\left(t\right)}\sum_{\beta=1}^{N(t)}\sum_{i=1}^{L} J_i\left(\mathbf{S}^{\alpha},\mathbf{S}^{\beta}\right) S_i^{\alpha} S_i^{\beta} - \mu N\left(t\right), \tag{8.10}$$

where $\mathbf{J}$ is the coupling strength associated with the given pair of positions in the genotype space, and μ is the set average population size, or the sustainable population size that is supported by the environment being considered. If we introduce the genotype space position occupancy, $n(\mathbf{S}, t)$, then the sum over the other individuals in Eq. (8.10) can be expressed more conveniently:

$$H\left(\mathbf{S}^{\alpha},t\right)=\frac{1}{N(t)}\sum_{\mathbf{S}}n\left(\mathbf{S},t\right)\sum_{i=1}^{L}J_{i}\left(\mathbf{S}^{\alpha},\mathbf{S}\right)S_{i}^{\alpha}S_{i}-\mu N(t).$$

The coupling strengths, $\mathbf{J}^{ab} \neq \mathbf{J}^{ba}$, can be chosen from a uniform distribution on $[-c, c]$, given some constant c. The sign of the coupling strength for a given pair of genotypes accounts for whether the directional interaction between them enhances (*e.g.*, food) or decreases (*e.g.*, predation) the reproduction of the given genotype. Mutualistic interactions have positive strengths, and competitive ones have negative values for both directions that constitute the interaction. As formulated (Christensen et al. 2002), the model disregards self-interaction of the genotypes ($\mathbf{J}(\mathbf{S}, \mathbf{S}) = \mathbf{0}$). Notably, setting $J_i(\mathbf{S}^a, \mathbf{S}^b) = J_0$ irrespective of $\mathbf{S}^a$, $\mathbf{S}^b$ results in the population that is distributed widely through the genotype space, whereas the variation of coupling strengths leads to its clustering, as described below. This observation, made in the cited work, is presumably consistent with the degree of clustering that we observed in Chap. 3 when considering ecologically neutral sympatric speciation (Higgs and Derrida 1992), clustering being stronger in the ecologically diversified case.

Reproduction occurs with probability, per unit time, that is proportional to

$$p_{\mathrm{off}}\left(\mathbf{S}^{\alpha},t\right)=\frac{\exp\left[H\left(\mathbf{S}^{\alpha},t\right)\right]}{1+\exp\left[H\left(\mathbf{S}^{\alpha},t\right)\right]}\in[0,1]$$

in the asexual case. For sexual reproduction, individual $\mathbf{S}^{\alpha}$ is selected at random and paired with individual $\mathbf{S}^{\beta}$, which is selected randomly insofar as the genetic distance between the two individuals is below the reproductive cut-off:

$$d=\frac{1}{2}\sum_{i=1}^{L}\left|S_{i}^{\alpha}-S_{i}^{\beta}\right|\le d_{\max}.$$

The pair is successful in producing offspring with probability

$$\sqrt{p_{\mathrm{off}}\left(\mathbf{S}^{\alpha},t\right)p_{\mathrm{off}}\left(\mathbf{S}^{\beta},t\right)}.$$

Each allele in the offspring is selected at random between the corresponding alleles of the parents. These conditions, once again, are analogous to the conditions of sympatric speciation that we discussed in Chap. 3. Mutation occurs with a specified probability per locus of the offspring. To offset the reproduction, an individual has a set probability, p_{kill}, per time step, to be annihilated. The time $N(\mathrm{t})/p_{\mathrm{kill}}$ is considered to be one generation.

It can be noted (Christensen et al. 2002) that, neglecting mutations, the average $n(\mathbf{S}, t)$, if it is not zero, must be equal to the average $p_{\mathrm{off}}(\mathbf{S}, t)/p_{\mathrm{kill}}$. Simulations show that the system can find configurations in the genotype space such that this equality is approximately satisfied for prolonged periods of time. However, an imbalance arising at some sites in this space can propagate through the system, leading to its global reconfiguration before the next quasi-steady state is found. Owing to the fact that the stable configurations represent alternative interaction strategies for the system of co-evolving genotypes, the quasi-steady states have been identified with evolutionarily stable strategies, as known from game-theoretical ecological studies (Christensen et al. 2002; Jensen 2004). Interestingly, like in the Higgs-Derrida model of sympatric speciation and the quasi-species formalism that were treated in Chap. 3, there is a critical mutation rate in the tangled-nature representation, beyond which the clustering in the genotype space does not occur (di Collobiano et al. 2003).

The evolved interaction network structure of the population of genotypes in the tangled-nature model can be seen in Fig. 8.1. The obvious difference of the genotype-based treatment from the cited species-based approach is in the large number of distinct nodes in the interaction network, which correspond to populated nodes in the genotype space. However, these nodes are clustered through the process broadly analogous to the one analyzed in Chap. 3 for species emergence in the neutral-evolution case. Keeping only the highly populated nodes, we can arrive at an interaction network whose structure is more familiar from real ecosystems analyzed on the species level (Fig. 8.1). The collective structure of the interactions—which can be understood as a food web even though the formalism of coupling weights captures other ecological interaction types equally well—displays a definite scaling with the number of species that are present at any given time (Fig. 8.2). However, quasi-stable states of the system display tighter distributions than transient states (Fig. 8.2). Periods of relative stability in the genotype distribution are punctuated by comparatively discrete transition periods (Fig. 8.3). Overall, we

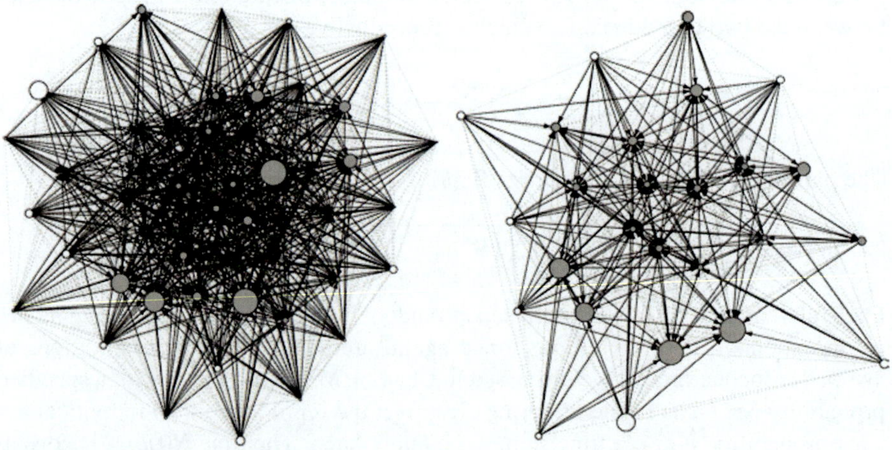

Fig. 8.1 Ecological connections between genotypes in the tangled-nature model. *Left*, full network evolved in a simulation, *right*, nodes with occupancy exceeding 10. Reproduced from Brinck and Jensen (2017). Reprinted from Journal of Theoretical Biology, Vol. 428, Katharina Brinck, Henrik Jeldtoft Jensen, The evolution of ecosystem ascendency in a complex systems based model, 18–25, copyright 2017, with permission from Elsevier

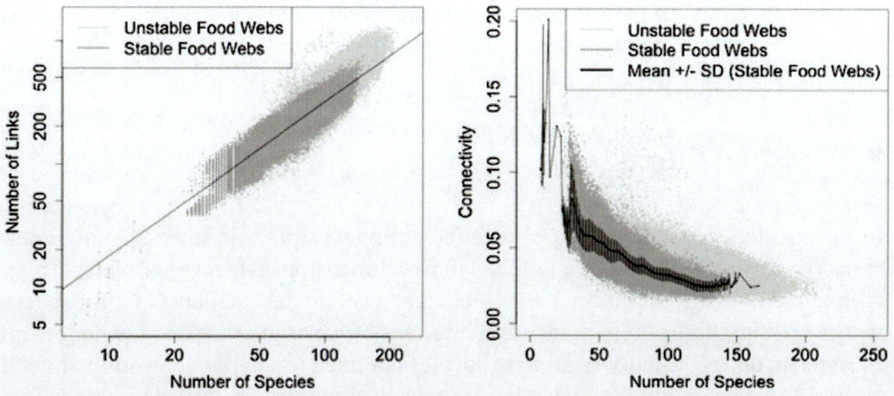

Fig. 8.2 Statistics of connectivity in the tangled-nature model. Reproduced from Brinck and Jensen (2017). Reprinted from Journal of Theoretical Biology, Vol. 428, Katharina Brinck, Henrik Jeldtoft Jensen, The evolution of ecosystem ascendency in a complex systems based model, 18–25, copyright 2017, with permission from Elsevier

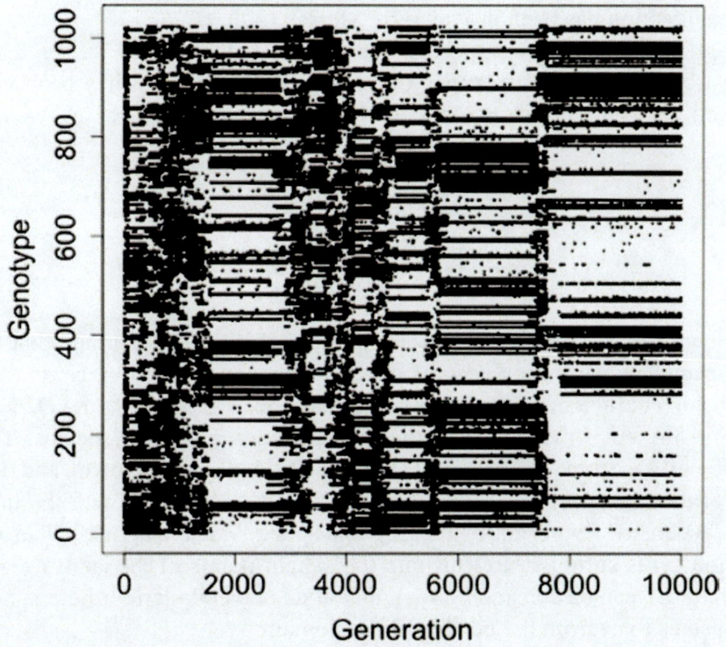

Fig. 8.3 Dynamics of genotype space occupancy distribution in the tangled-nature model. Reproduced from Brinck and Jensen (2017). Reprinted from Journal of Theoretical Biology, Vol. 428, Katharina Brinck, Henrik Jeldtoft Jensen, The evolution of ecosystem ascendency in a complex systems based model, 18–25, copyright 2017, with permission from Elsevier

observe that temporal pattern formation is a general feature in ecological dynamics arising from species interactions.

8.3 Spatial Effects

In the physical environment, species interaction takes place in three spatial dimensions. These can be effectively reduced to two dimensions when geographical (planetographical) distributions are concerned, and also, on the smaller scale, for species whose ecology limits them to medium interfaces such as the surface of water or soil. In some instances, usually relating to species confined to a certain elevation or depth around a surface feature (an island, a lake, or a mountain), the effective dimensionality may be one. The two-dimensional description is frequently of most interest—as we ourselves dwell in this dimensionality—and represents the environment dimensionality for a large number of advanced life forms. Distribution of species in their physical environment that is shaped by the species' interactions is one of the classical applications of quantitative analysis in biology. A broad account of the modeling approaches, empirical data collection, and methodological questions of their juxtaposition has been provided by Murray (2003).

Spatial treatment of two-species interactions belonging to the predator-prey, mutualistic, and competitive types of ecological relationship has been comparatively well researched (see Owolabi and Patidar 2016). In these cases, we have

$$\begin{aligned} \frac{\partial u}{\partial t} &= D_u \frac{\partial^2 u}{\partial x^2} + f(u,v), \\ \frac{\partial v}{\partial t} &= D_v \frac{\partial^2 v}{\partial x^2} + g(u,v) \end{aligned} \tag{8.11}$$

for the local spatial densities of species u and v, if it can be assumed that the individual species redistribution accepts the diffusion approximation. Here, the apparent diffusion coefficients of the two species have been denoted D_u and D_v. Functions f and g in Eq. (8.11) encapsulate the local dynamics of each species. The local dynamics arises from the species' own density-dependent growth and from the effects of the interaction with the other species, in accordance with the other species' local density. Representing the physical space available to the species with one dimension (x) is sufficient to illustrate the salient points of the analysis. A closed interval, $(0, L)$, can be considered such that at its two boundaries, there is no flux of either species to or from the outside of the domain:

$$\left.\frac{\partial u}{\partial x}\right|_{x=0,\ L} = 0, \quad \left.\frac{\partial v}{\partial x}\right|_{x=0,\ L} = 0.$$

To determine the system behavior, it is expedient to consider the dynamics of the chosen form of species interaction (*f*, *g*) in the absence of species migration. Then we have

$$\frac{du}{dt} = f(u,v), \qquad \frac{dv}{dt} = g(u,v). \tag{8.12}$$

The spatially homogeneous system (Eq. 8.12), for realistic parameter choices, will generally have a stable solution, $(\hat{u},\hat{v})$. Introducing the Jacobian matrix,

$$J = \begin{pmatrix} J_{11} & J_{12} \\ J_{21} & J_{22} \end{pmatrix} = \begin{pmatrix} \left.\frac{\partial f}{\partial u}\right|_{\hat{u},\hat{v}} & \left.\frac{\partial f}{\partial v}\right|_{\hat{u},\hat{v}} \\ \left.\frac{\partial g}{\partial u}\right|_{\hat{u},\hat{v}} & \left.\frac{\partial g}{\partial v}\right|_{\hat{u},\hat{v}} \end{pmatrix}, \tag{8.13}$$

and considering the characteristic equation,

$$\lambda - \mathrm{tr}J\lambda + \det J = 0,$$

we observe that

$$\mathrm{tr}J = J_{11} + J_{22} \langle 0, \qquad \det J = J_{11}J_{22} - J_{12}J_{21} \rangle 0. \tag{8.14}$$

The modes cos(*kx*) of a small perturbation to the homogeneous solution will grow with exponential rates λ_k. After introducing this notation, the condition for the Turing diffusional instability of the homogeneous state is

$$\begin{gathered} k^4 + \Psi_1 k^2 + \Psi_2 = 0, \\ \Psi_1 = -\left(\frac{J_{11}}{D_u} + \frac{J_{22}}{D_v}\right), \qquad \Psi_2 = \frac{\det J}{D_u D_v}. \end{gathered} \tag{8.15}$$

Positivity of the roots of Eq. (8.15) requires that $\Psi_1 < 0$, *i.e.*, $D_v J_{11} + D_u J_{22} > 0$. Considering Eq. (8.14), we must have $D_v \neq D_u$ and J_{11}, J_{22} that are of opposite sign. The Jacobian (Eq. 8.13), therefore, may be of one of the two types:

$$J = \begin{pmatrix} + & - \\ + & - \end{pmatrix}, \qquad J = \begin{pmatrix} + & + \\ - & - \end{pmatrix}.$$

In either case, the above condition on the diffusion coefficients stipulates that $|J_{22}| > J_{11}$, while $J_{12}J_{21} < 0$. Moreover, $D_v/D_u > J_{22}/J_{11} > 1$. Thus, the activator species must diffuse more slowly, compared with the inhibitor species. Further examining Eq. (8.15), we can note the second condition for instability of the homogeneous solution $(\hat{u},\hat{v})$ in the presence of species migration: $\Psi_1^2 - 4\Psi_2 > 0$, *i.e.*, $(D_v J_{11} + D_u J_{22})^2 > 4 D_u D_v \det J$. The unstable modes correspond to

$$\frac{-\Psi_1-\sqrt{\Psi_1^2-4\Psi_2}}{2}<k^2<\frac{-\Psi_1+\sqrt{\Psi_1^2-4\Psi_2}}{2}.$$

Transitioning now to specific examples, we can consider the following equations for a spatially distributed, comparatively realistic predator-prey system (see Murray 2002; Garvie 2007; Owolabi and Patidar 2016):

$$\begin{aligned}\frac{\partial U}{\partial T}&=D_1\frac{\partial^2 U}{\partial X^2}+U\left[\alpha\left(1-\frac{U}{K}\right)-\frac{\gamma V}{U+\delta}\right],\\ \frac{\partial V}{\partial T}&=D_2\frac{\partial^2 V}{\partial X^2}+V\beta\left(1-\frac{hV}{U}\right).\end{aligned} \tag{8.16}$$

U here is the density of the prey, and V, of the predator. The positive parameters K, α, β, γ in Eq. (8.16) describe the two species' logistic (intrinsically limited) proliferation, which distinguishes this modern formulation of a predator-prey model from the elementary Lotka-Volterra approach in the last section. The system nondimensionalizes to the general form (Eq. 8.11) with

$$f(u,v)=u(1-u)-\frac{\mu u}{u+\varphi}v,\quad g(u,v)=\psi v-\frac{\psi v^2}{u},$$

$$u=\frac{U}{K},\quad v=\frac{hV}{K},\quad t=\alpha T,\quad \mu=\frac{\gamma}{h\alpha},\quad \psi=\frac{\delta}{K},\quad D_u=1,\quad D_v=\frac{D_2}{D_1}.$$

In the case of a competitive interaction and, in fact, in any instance of mutual suppression between the species, a spatial formalization that can be considered is

$$\begin{aligned}\frac{\partial U}{\partial T}&=D_1\frac{\partial^2 U}{\partial X^2}+\alpha_1 U\left(1-\frac{U}{K_1}-\beta_1\frac{V}{K_1}\right),\\ \frac{\partial V}{\partial T}&=D_2\frac{\partial^2 V}{\partial X^2}+\alpha_2 V\left(1-\frac{V}{K_2}-\beta_2\frac{U}{K_2}\right),\end{aligned} \tag{8.17}$$

which nondimensionalizes to the form of Eq. (8.11) with

$$f(u,v)=u-u^2-\varphi uv,\quad g(u,v)=\mu\left(v-v^2-\varphi uv\right),$$

$$u=\frac{U}{K_1},\quad v=\frac{V}{K_2},\quad t=\alpha_1 T,\quad \mu=\frac{\alpha_2}{\alpha_1},\quad \varphi=\frac{\beta_1 K_2}{\beta_2 K_1},\quad D_u=1,\quad D_v=\frac{D_2}{D_1}.$$

As a final example, for a spatially distributed mutualistic system, in which each species locally enhances the growth of the other, we can adopt the same form as in Eq. (8.17), but select negative β_1 and β_2 to reflect the fact that the other species' density has a beneficial effect on the given species' growth. Species density patterns that can arise and propagate in a two-dimensional space populated with two species,

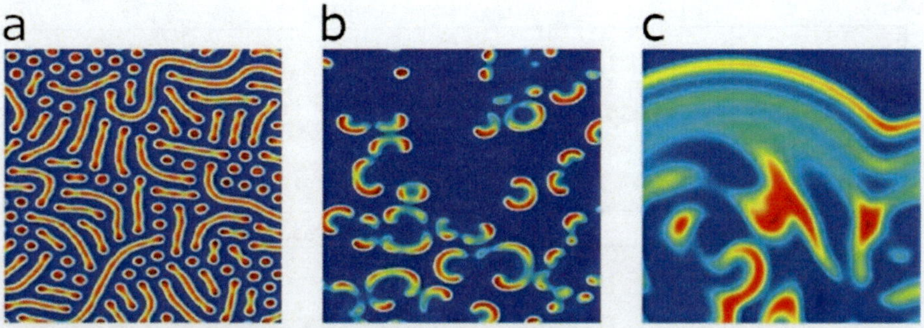

Fig. 8.4 Two-dimensional patterns occurring with species interaction. (**a**) Mutualistic relationship, (**b**) competitive relationship, (**c**) predator-prey relationship (predator density shown). Reproduced in a modified form, with permission, from Owolabi and Patidar (2016)

whose interaction comports with the introduced predator-prey, competitive, and mutualistic formalisms, are shown in Fig. 8.4. Qualitative and detailed comparison of these potentialities with distributions observed in biogeographic studies represents a fundamentally important direction of research.

Spatial distribution is a significant factor affecting evolution. The relative contributions of population-intrinsic mechanisms of speciation (such as those that we have considered, as the most fundamental, in Chap. 3), of species interactions (Sect. 8.2), and of spatial separation have been recently systematized by Barraclough (2019). Parapatric and allopatric speciation, which occurs in geographically isolated—whether partially or fully—populations has received considerable attention in evolutionary research. An extension of the Higgs-Derrida framework that we introduced in Chap. 3 has been developed to address the interplay of the population-intrinsic reproductive isolation mechanisms with geographic isolation and migration in such cases (Manzo and Peliti 1994).

Another, presumably more commonly occurring, situation has become possible to analyze using the topopatric approach (de Aguiar et al. 2009). In it, reproduction is dependent on distance in contiguous physical space analogously to its dependence on distance in genomic space. This model is a spatial extension of the finite-genome version of the Higgs-Derrida model (Chap. 3). Genetically isolated species that arise in this model occupy spontaneously organized domains in the physical space that is available to the metapopulation (Fig. 8.5). Statistical properties of the spatially self-organized species in the topopatric model follow a number of quantitative laws that are observed in real species (de Aguiar et al. 2009). These laws represent both a generalization and a concretization of statistical laws of neutral speciation (Hubbell 2001; Banavar and Maritan 2009).

Recently, it has been shown that geographical (planetographical) separation via the self-organized topopatric mechanism permits speciation with a smaller number of genes that are involved in reproductive isolation, compared with the number required for speciation in strict sympatry, *i.e.*, when spatial reproductive mixing is somehow ensured (de Aguiar 2017). In a parapatric situation, moreover, a smaller

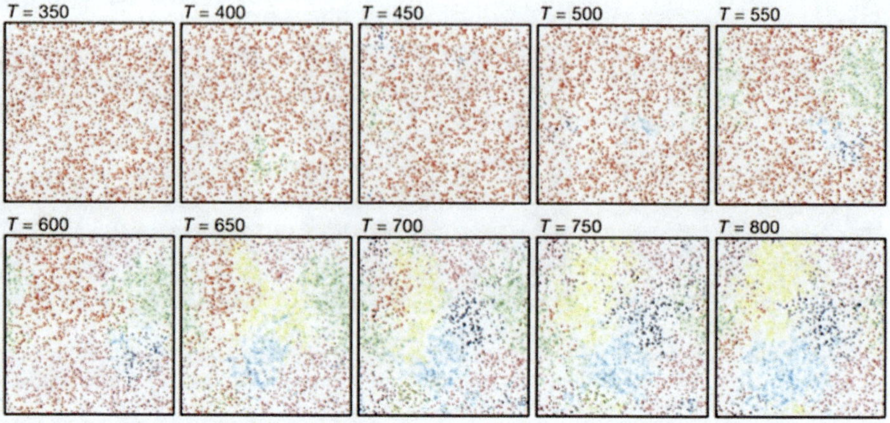

Fig. 8.5 Formation of spatially separated species in the topopatric model. Reproductively isolated species are color-coded. Reproduced from de Aguiar et al. (2009). Reprinted by permission from Springer Nature Customer Service Centre GmbH: Springer Nature, Nature (https://www.nature.com/), Global patterns of speciation and diversity, M. A. M. de Aguiar et al., copyright 2009

number of such genes makes speciation more likely to occur (Yamaguchi and Iwasa 2016). The topopatric mechanisms have also been studied in application to sexual and diploid organisms (Baptestini et al. 2013; Schneider et al. 2016). Most recent work has addressed development of algorithms for tracing evolutionary history in individual-based simulations and deriving the corresponding phylogeny (Costa et al. 2018). These methods make it possible to determine the impact of topopatric mechanisms of speciation on macroevolution, which represents an exciting direction for further application of the spatial formalism.

References

Bak P, Sneppen K (1993) Punctuated equilibrium and criticality in a simple model of evolution. Phys Rev Lett 71:4083

Banavar J, Maritan A (2009) Towards a theory of biodiversity. Nature 460:334–335

Baptestini EM, de Aguiar MAM, Bar-Yam Y (2013) Conditions for neutral speciation via isolation by distance. J Theor Biol 335:51–56

Barraclough TC (2019) The evolutionary biology of species. Oxford University Press, Oxford

Brinck K, Jensen HJ (2017) The evolution of ecosystem ascendency in a complex systems based model. J Theor Biol 428:18–25

Christensen K, di Collobiano SA, Hall M et al (2002) Tangled nature: a model of evolutionary ecology. J Theor Biol 216:73–84

Costa CLN, Marquittib FMD, Perez I et al (2018) Registering the evolutionary history in individual-based models of speciation. Physica A 510:1–14

de Aguiar MAM (2017) Speciation in the Derrida–Higgs model with finite genomes and spatial populations. J Phys A 50:085602

de Aguiar MAM, Baranger M, Baptestini EM et al (2009) Global patterns of speciation and diversity. Nature 460:384–387 di Collobiano SA, Christensen K, Jensen HJ (2003) The tangled nature model as an evolving quasi-species model. J Phys A 36:883–891

Garvie MR (2007) Finite-difference schemes for reaction-diffusion equations modeling predator-prey interactions in MATLAB. Bull Math Biol 69:931–956

Higgs PG, Derrida B (1992) Genetic distance and species formation in evolving populations. J Mol Evol 35:454–465

Hubbell SP (2001) The unified neutral theory of biodiversity and biogeography. Princeton University Press, Princeton

Jensen HJ (2004) Emergence of species and punctuated equilibrium in the tangle nature model of biological evolution. Phys A 340:697–704

Kauffman SA, Johnsen S (1991) Coevolution to the edge of chaos: coupled fitness landscapes, poised states, and coevolutionary avalanches. J Theor Biol 149:467–505

Lotka AJ (1920) Undamped oscillations derived from the law of mass action. J Am Chem Soc 42:1595–1599

Lotka AJ (1925) Elements of physical biology. Williams and Wilkins, Baltimore

Manzo F, Peliti L (1994) Geographic speciation in the Derrida-Higgs model of species formation. J Phys A 27:7079–7086

May RM (1975) Stability and complexity in model ecosystems. Princeton University Press, Princeton, NJ

Murase Y, Rikvold PA (2018) Conservation of population size is required for self-organized criticality in evolution models. New J Phys 20:083023

Murray JD (2002) Mathematical biology I: an introduction. Springer, New York

Murray JD (2003) Mathematical biology II: spatial models and biomedical applications. Springer, New York

Owolabi KM, Patidar KC (2016) Numerical simulations of multicomponent ecological models with adaptive methods. Theor Biol Med Model 13:1–25

Schneider DM, Baptestini EM, de Aguiar MAM (2016) Diploid versus haploid models of neutral speciation. J Biol Phys 42:235–245

Solé RV, Manrubia SC (1996) Extinction and self-organized criticality in a model of large-scale evolution. Phys Rev E 54:R42–R45

Volterra V (1926) Variazioni e fluttuazioni del numero d'individui in specie animali conviventi. Mem Acad Lincei 2:31–113

Whittaker RH (1975) Communities and ecosystems. Macmillan, New York

Yamaguchi R, Iwasa Y (2016) Smallness of the number of incompatibility loci can facilitate parapatric speciation. J Theor Biol 405:36–45

Chapter 9
Summary and Outlook

In the preceding chapters, we have presented a selection of general biological topics whose analysis has benefited from application of closely related dynamical systems methods. Biomass kinetics allowed us to introduce all readers to the mathematical and conceptual apparatus of dynamical systems theory on simple examples of steady states, their stability and instability, as well as oscillations and apparent chaos. Speciation was considered, at first, on the elementary level of species emergence in neutral evolution of well-mixed and homogeneous populations, along with conditions for species maintenance that relies on mutation-selection equilibria. Biological control mechanisms provided material on which to elaborate the analysis of oscillatory behavior, as applied to both environmental signal processing and control of effectors such as muscles involved in breathing. Switch-like responses of the cell physiology to gradual changes in the environment could be treated using related analytical methods. When considering molecular self-organization in the cell, we encountered somewhat different types of dynamics that captured spatial distributions shaped by diffusional and non-diffusional motion, as well as by apparent motion that arises from linear polymerization. Nevertheless, the introduced apparatus of differential equations and steady states served the purpose of this analysis as well. Cell mechanics presented yet another variation on the theme, where we focused on stability of mechanically static equilibria as the main approach to explain how the cell makes the choice between the developmental fates linked to symmetric and asymmetric division. The recently elucidated problems of morphogenesis illustrated an extension of the method of spatially distributed dynamical systems to the multicellular scale, where spatiotemporal periodicity is observed in the patterns of plant growth and animal development. Finally, we revisited speciation and biomass kinetics from the standpoint of species interaction, and analyzed the spatiotemporal patterns that arise in ecology and coevolution.

Throughout, the emphasis was on laying out the illustrative examples in sufficient detail for the reader to follow the logic of each application of the dynamical systems method. Inevitably, in the space of this monograph, the coverage of topics

I. Maly, *Quantitative Elements of General Biology*,
https://doi.org/10.1007/978-3-030-79146-9_9

is closer to a sketch of the quantitative elements of general biology than to a compendium. Furthermore, although it is hoped that the universality of the dynamical systems approach has been sufficiently brought out, the treatment differs in many ways between the applications. In certain instances, the formalism can be unified without much difficulty. The exposition here does that in some measure, but keeps the focus on the intrinsic logic of the selected applications, viewing it as more valuable to the reader at this stage in the development of unified biological dynamical systems theory.

Beginning with the relatively simple approaches to biomass growth (Chap. 2), we have encountered several fundamental principles in behavior of dynamical systems. By considering single-variable (scalar) descriptions of population sizes, a range of well-known cases of population dynamics (see also Murray 2002) have been laid out, which exhibited stable equilibria, unstable equilibria, and switching between these states with the change in parameters, or with external forcing of the system. Although mathematically simple, the models considered were applicable to both asexual and sexual reproduction, insofar as the population was genetically uniform and spatially well-mixed. The principle of domains of attraction toward stable equilibria being separated by unstable equilibria in the parameter space that was illustrated on these examples illuminates behavior of more complex systems. The second general effect evident in the scalar biomass models is hysteresis—dependence of the state of the system on its history, including the history of any parameter variations, and the resulting possibility that the system may follow different trajectories, depending on the direction of parameter change.

Generational dynamics of the simple populations can remain univariate in its description of the population size at any given time, but even in the case of non-overlapping generations, the population's trajectory through time resembles multivariate (for example, bivariate) population models because of the dependence of the current state on the point occupied by the system in the past. This gives rise to richer dynamics than possible in the continuous-time biomass growth, particularly to oscillations. In addition to the stability and instability criteria in the discrete-time case, we have considered bifurcations of the steady-state population size into cycles of two or more repeated values. The nature of aperiodic and extremely long-period time sequences that can be exhibited by populations with generations was also considered, leading to the concept of deterministic chaos.

Selection-mutation equilibria form the basis for quantitative characteristics of all other biological processes due to the fact that the latter are controlled (whether directly or indirectly) by genetically determined parameters. In Chap. 3, we considered the governing principles behind steady-state distributions in genotype space. The considered selection-mutation models explain why the steady-state frequency of the selectively favored alleles can exhibit a strongly nonlinear response to the relative power of the two processes, even when each of them has only a weak impact on the generational time scale. One of the seminal analyses in this area is quasi-species theory (Eigen et al. 1988). In our exposition, following Higgs (1994), the original quasi-species treatment is but a special case characterized by a step function form of fitness in the genomic space. Nevertheless, the principles of quasi-species, whereby a reproductively cohesive population, or biological species, exists as a well-defined but broad cloud of mutational variants, and nonlinear transitions take place in the shape of this distribution in response to parameter change, prove applicable to other fitness landscapes. Moreover, as the material in subsequent chapters demonstrates, these principles are illuminating of biological dynamics well beyond genetic equilibria.

The foundational models of mutation-selection equilibria presuppose reproductive cohesion of the population, and as such are applicable only to established species. The very existence of species, however, is an essential attribute of living matter that requires a quantitative explanation. The apparatus for analyzing the elementary dynamics behind species emergence is closely related to that for mutation-selection equilibria and entails examining the consequences of genetically assortative mating (Sect. 3.2). As we have seen, spontaneous partitioning of the population leads to formation of reproductively cohesive and reproductively isolated subsets. The biological process addressed by this approach is sympatric speciation, which is intrinsic to the population's dynamics in the genotype space and occurs in the absence of spatial separation of the subpopulations. By comparison, parapatric speciation—especially its recently characterized topopatric mechanism—involves an interplay of genetic and spatial biological dynamics that collectively explain more than the intrinsic tendency of living matter toward species formation. Because of the comparative complexity and derivative nature of such combined genetic-spatial evolutionary mechanisms, we set aside their treatment till the last chapter. By maintaining the focus on the most elementary in Chap. 3, a new degree of formalism unification between mutation-selection equilibria and sympatric speciation has been achieved. More work, however, is needed toward this goal. A promising direction for elaboration of findings pertaining to the mechanism of speciation is adaptation of the general approach to concrete systems studied in molecular and organismal detail (Gavrilets 2014), in order to better understand how the general mechanism operates. The impact of the gene interaction network structure, and specific physiology affecting mating, on emergence of reproductive isolation may be of particular interest in this connection.

Alongside biomass kinetics and speciation, homeostasis mechanisms belong to the category of the most elemental biological processes, because they are responsible for maintaining the physical distinction between living and non-living matter. In Chap. 4 we considered biological mechanisms of control and the closely related mechanisms of signal processing, which collectively maintain constancy of the organisms' internal environment relative to their variable external environment, and effect the organisms' adaptation to the latter. In addition to gradual control, cells and organisms exhibit all-or-nothing responses to external signals and transitions between qualitatively different control regimes in response to variation of the physiological parameters. We have considered all-or-nothing responses in intracellular kinase cascades, and also on the cellular level, using the example of membrane action potential (Hodgkin and Huxley 1952). Integrative extensions of the action potential model to the level of neuronal networks controlling the breathing cycle have been presented, and the cyclic dynamics was additionally illustrated on the examples of the kinase cascades and calcium signaling. The same general form of signaling kinetics, it was shown, can support switch-like behavior in the molecular interaction circuit. Using the latest results concerning the paradigmatic *lac* operon (Jacob et al. 1960), we have seen how switches can operate in molecular networks involving gene expression regulation. Cyclicity and hysteresis, which had been already encountered in biomass kinetics, and which underlie the behavior of the control systems, demonstrated the continuity of key types of dynamical behavior on different levels of biological organization. Along the way, some clarification of the physical description of action potential, and a certain easily achievable degree of unification of the action-potential and integrative formalisms, as well as across the different applications of kinase cascade kinetics, have been offered. Much scope remains for future work in this direction. Flexibility of control and information transmission circuits and their capacity to produce qualitatively different dynamic behavior and responses also present interest at the intersection with the general biological problem of cell heterogeneity and individuality. It may be expected that significant insights into how the diverse dynamic repertoire is supported by conserved molecular network motifs, such as the kinase cascades or calcium homeostasis system, will come from analyzing individual cell responses through the prism of kinetic models—for example, using Bayesian inference and clustering in the space of the inferred parameters (cf. Yao et al. 2016; Tilūnaitė et al. 2017; Haggerty and Purvis 2021).

The physical-chemical principles in the organization of the cell as the nearly universal unit of living matter, and the dynamical system effects arising from the interaction of molecules responsible for this organization, are among the central problems in general biology. Whereas the chemical nature of such molecular components may be difficult to generalize from known life forms to those that remain to be discovered beyond Earth, the expectation to find cellular organization to be widespread in the universe can serve as a base assumption in the development of biology that is generalizable from the singular example presented to us by life on this planet. Apart from the breakdown of living matter into cells, the common principles that can be generalized from terrestrial cells include the many cells' internal organization exhibiting a focal center, anisotropy and local directionality of their cytoplasm, and overall cell polarity. As reviewed in Chap. 5, these arise from the presence of cytoskeletal fibers, their intrinsic macromolecular-level polarity and kinetics, and from the constraints imposed on this kinetics by the cell boundary, which is a fundamental factor in the heritable cell structure (Harold 2001). Accordingly, we have treated the spatial kinetics of self-organization of the cytoskeleton structure with a focal point in the cell center, and the kinetic genesis of the orientational order in cytoskeletal filaments underlying the cell boundary itself. Further, the signaling formalism from the preceding chapter was generalized to three spatial dimensions, where it is capable of capturing self-organization of signaling that defines spontaneously emerging, distinct poles of the cell.

The instabilities of the homogeneous state in the cytoskeletal and signaling examples presented our first instances of Turing-type spatial instabilities (Turing 1952), while the dynamics of cytoskeletal filaments associated with the cell boundary exhibited features already familiar from the quasi-species theory. Thus, although the mathematical form of some of the models in Chap. 5 was more complex than is typical in biological applications of distributed dynamical systems, due to the intrinsic polarity of the molecular components involved, the core features of these processes' behavior are shared across multiple levels of biological organization.

The ongoing work on building integrated dynamical models of spatial self-organization of cytoskeleton and cell signaling molecules has been briefly reviewed in Chap. 5. Advances in this area can add mechanistic depth to our understanding of the cell form, particularly cell polarity. They can also yield significant insights into the interplay of self-organization effects that are intrinsic, respectively, to cell signaling and the cytoskeleton. A related direction of study, where recent work has held out promise of a fundamental breakthrough, concerns mechanisms of cell division. Of particular interest here is the problem of coordinated progression of the signaling and control system, on the one hand, and cell growth and metabolism, on the other, through the cell cycle (Surovtsev et al. 2007; Macía and Solé 2007a, b; Surovtsev et al. 2009; Bigan et al. 2016).

Systems biomechanics of the cell is at the root of morphogenesis and motility that are not meaningfully reducible to even the complex self-organization effects in the molecular realm (Maly 2013), but arise entirely as properties of the structure—primarily, the cell boundary confining the cytoplasmic material inside—which, in developed life, is only inherited from cell to cell (Harold 2001). In Chap. 6, we sketched out the most elementary effects in systems biomechanics of the cell, along with some approaches to their analysis, on the sufficiently investigated example of this type of dynamics, which is the microtubule cytoskeleton confined in the cell boundary. The fundamental result concerning this system is that the forces resulting from the aster's confinement induce asymmetry in the microtubule cytoskeleton. A special but biologically essential case is asymmetry induction in the mitotic cytoskeleton, which leads to asymmetric cell division. We also considered how the direction of the microtubule aster asymmetry is selected relative to cell-cell contacts. Analysis shows that the physics of the confined microtubule cytoskeleton exhibits a series of system-level effects, which include, in addition to the spontaneous symmetry-breaking, irreversibility, structural multistability, and multiperiodic movement. These effects extend to the realm of cell mechanics the observations made in the preceding chapters about the biochemical and cell-physiological dynamics. Furthermore, systems biomechanics of the cell supplies the foundation for a hierarchy of irreversible and periodic developmental mechanisms that operate on the multicellular level, some of which are the subject of the subsequent chapter. Future studies in systems biomechanics of the cell, therefore, should be able to close the gap between our understanding of self-organization on the macromolecular and tissue levels.

Considering multicellular morphogenesis (Chap. 7), we have noted a remarkable progress in recent years that has led to productive merging of classical pattern formation theory with systems and molecular experiments in a number of areas, and to a synthesis of biomechanical and biochemical approaches. So, for example, quantitative treatment of apical meristem instability generating the phyllotactic patterns has integrated the effects of tunica buckling and auxin transport mechanisms, pinpointing conditions that lead to their cooperation. The most interesting among these conditions, which supply a framework for experimental analysis, is the regime where cooperation occurs despite different patterns being favored by the two mechanisms. Derivation of approximations to the mechanochemical formalism also represents a significant advancement due to the in-depth quantitative comparison between fundamental theory and empirical data that it has made possible. In the field of embryo segmentation, by comparison, the latest results suggest a unification of the sequential and simultaneous segmentation mechanisms as two regimes that can be selected by evolution in a single basic system dynamics. The selected examples demonstrate that the dynamical systems analysis in this area has revealed quantitative effects that extend the observations made in the molecular and cellular systems to the multicellular domain (instability, oscillations, and spatial patterning). One of the most fundamentally important directions is investigation of how evolvability determines the patterning mechanisms. Open questions concern both the molecular architecture of the developmental networks and the temporal program underlying the organism's developmental plan (see, e.g., Vroomans et al. 2018; Landge et al. 2020).

Our survey of the elementary quantitative phenomena in general biology concluded with a classical area of application for the dynamical systems method: interactions between species in the biogeographical and evolutionary domain (Chap. 8). After reviewing the origin of stability, instability, and oscillations arising from species interaction, we considered the structure of evolutionary ecological representations that exhibit a more complex temporal dynamics, which is reminiscent of punctuated equilibrium. Self-organized criticality in this instance leads to avalanches of reorganization of the species' ecological relationships as well as occupancy of the genotype space; once the rapidly evolving system finds a new equilibrium, the relative stasis can persist for a long time. A spatial version of the general two-species ecological formalism was also analyzed, including the conditions for the Turing instability arising from species migration and three applications to ecological interaction types (mutualism, predation, and competition). Concluding our survey was a recently developed spatial evolutionary formalism, which demonstrates facilitation of speciation by biogeographical (bioplanetographical) distribution even in the absence of any pre-existing boundaries between populations. Boundaries of spatial domains occupied by species in this type of dynamics arise from the system's self-organization in the physical space simultaneously with its self-organization in the genomic space. The dynamics considered in Chap. 8 elaborate on the elementary biomass and evolutionary effects with which we started our survey, by re-examining these processes in a more realistic setting for developed life, which is characterized by species interaction and competition for space.

An interesting direction for extending the findings concerning the dynamics of evolving ecological systems is evolution of inter-organism communications systems (see Segovia-Martín and Balari 2020). This area of application and further development of the principles of dynamic self-organization includes also evolution of the informational content that exists on the biological substrate of the communication systems. Similarly, planetary dynamics, particularly planetary chemistry and climate, form a system that incorporates the activity of the biosphere as one of its determinative factors (see Yamagishi et al. 2019). Self-organization on the planetary scale can be studied not only in the context of emergence of conditions for life and planetographic signatures of past and extant life, but also with respect to the role of the biosphere in planetary evolution.

Applicability of related mathematical methods of dynamical systems theory to the diverse topics sketched out in Chap. 2–8 is, of course, not coincidental. It reflects the unity of laws of behavior of biological systems on the vastly different scales, from molecular to biogeographical. Much work remains to be done toward unification of biological theory around the commonality of system-level dynamics, expressible in a common mathematical language. It is hoped that the overview offered here can serve as a sufficiently broad sketch of such future theory, and provide an additional impetus for its development.

References

Bigan E, Paulevé L, Steyaert JM et al (2016) Necessary and sufficient conditions for protocell growth. J Math Biol 73:1627–1664

Eigen M, McCaskill J, Schuster P (1988) The molecular quasi-species. J Phys Chem 92:6881–6891

Gavrilets S (2014) Models of speciation: where are we now? J Hered 105:743–755

Haggerty RA, Purvis JE (2021) Inferring the structures of signaling motifs from paired dynamic traces of single cells. PLOS Comp Biol 17:e1008657

Harold FM (2001) The way of the cell. Oxford University Press, New York

Higgs PG (1994) Error thresholds and stationary mutant distributions in multi-locus diploid genetics models. Genet Res 63:63–78

Hodgkin AL, Huxley AF (1952) A quantitative description of membrane current and its application to conduction and excitation in nerve. J Physiol 117:500–544

Jacob F, Perrin D, Sanchez C et al (1960) Operon: a group of genes with the expression coordinated by an operator. C R Hebd Seances Acad Sci 250:1727–1729

Landge AN, Jordan BM, Diego X et al (2020) Pattern formation mechanisms of self-organizing reaction-diffusion systems. Dev Biol 460:2–11

Macía J, Solé RV (2007a) Protocell self-reproduction in a spatially extended metabolism-vesicle system. J Theor Biol 245:400–410

Macía J, Solé RV (2007b) Synthetic Turing protocells: vesicle self-reproduction through symmetry-breaking instabilities. Philos Trans R Soc Lond Ser B Biol Sci 362:1821–1829

Maly IV (2013) Systems biomechanics of the cell. Springer, Boston

Murray JD (2002) Mathematical biology I: an introduction. Springer, New York

Segovia-Martín J, Balari S (2020) Eco-evo-devo and iterated learning: towards an integrated approach in the light of niche construction. Biol Philos 35:42

Surovtsev IV, Morgan JJ, Lindahl PA (2007) Whole-cell modeling framework in which biochemical dynamics impact aspects of cellular geometry. J Theor Biol 244:154–166

Surovtsev IV, Zhang Z, Lindahl PA et al (2009) Mathematical modeling of a minimal protocell with coordinated growth and division. J Theor Biol 260:422–429

Tilūnaitė A, Croft W, Russell N et al (2017) A Bayesian approach to modelling heterogeneous calcium responses in cell populations. PLOS Comp Biol 13:e1005794

Turing AM (1952) The chemical basis of morphogenesis. Philos Trans R Soc Lond B 237:37–72

Vroomans RMA, Hogeweg P, Ten Tusscher KHWJ (2018) Around the clock: gradient shape and noise impact the evolution of oscillatory segmentation dynamics. EvoDevo 9:24
Yamagishi A, Kakegawa T, Usui T (eds) (2019) Astrobiology. Springer, Signapore
Yao J, Pilko A, Wollman R (2016) Distinct cellular states determine calcium signaling response. Mol Syst Biol 12:894

Appendix: Phase Plane Analysis

In biological applications, dynamics of molecular species, populations of organisms, etc., are described as systems of ordinary differential equations. Frequently, evaluating stability of these systems' steady states is the question of first importance. Mathematical methods for addressing this question through phase plane analysis are well elaborated (see, e.g., Guckenheimer and Holmes 1983, Jordan and Smith 1999), and practical overviews for biological applications are also available (e.g., Odell 1980, Murray 2002). The main results can be followed on the example of a simple system,

$$\frac{dx}{dt} = f(x,y), \qquad \frac{dy}{dt} = g(x,y). \tag{A.1}$$

Phase trajectories of Eq. (A.1) satisfy

$$\frac{dx}{dy} = \frac{f(x,y)}{g(x,y)} \tag{A.2}$$

and are unique at any point (x, y), except singular points (x_s, y_s) where

$$f(x_s,y_s) = g(x_s,y_s) = 0.$$

In the neighborhood of a singular point, taking (0, 0) for its coordinates, analytic f and g can be approximated linearly, yielding

$$\frac{dx}{dy} = \frac{ax+by}{cx+dy}, \qquad \mathrm{A} = \begin{pmatrix} a & b \\ c & d \end{pmatrix} = \begin{pmatrix} f_x & f_y \\ g_x & g_y \end{pmatrix}_{(0,0)}. \tag{A.3}$$

To obtain the local phase trajectories, we solve

I. Maly, *Quantitative Elements of General Biology*,
https://doi.org/10.1007/978-3-030-79146-9

$$\frac{dx}{dt} = ax + by, \quad \frac{dy}{dt} = cx + dy. \tag{A.4}$$

This can be done by considering the eigenvalues of A:

$$\begin{vmatrix} a-\lambda & b \\ c & d-\lambda \end{vmatrix} = 0 \quad \Rightarrow$$
$$\lambda_1, \lambda_2 = \frac{1}{2}\left(a + d \pm \left[(a+d)^2 - 4\det A\right]^{1/2}\right), \tag{A.5}$$

leading to

$$\begin{pmatrix} x \\ y \end{pmatrix} = c_1 \mathbf{v}_1 \exp(\lambda_1 t) + c_2 \mathbf{v}_2 \exp(\lambda_2 t) \quad (\lambda_1 \neq \lambda_2). \tag{A.6}$$

In Eq. (A.6), c_1 and c_2 are arbitrary constants and $\mathbf{v}_1$, $\mathbf{v}_2$ are the eigenvectors of A:

$$\mathbf{v}_i = \left(1 + p_i^2\right)^{-1/2} \begin{pmatrix} 1 \\ p_i \end{pmatrix}, \quad p_i = \frac{\lambda_i - a}{b} \quad (b \neq 0). \tag{A.7}$$

These solution forms imply $\lambda_1 \neq \lambda_2$, which can be expected in realistic models.

The possibilities concerning the behavior of the system in the neighborhood of the selected singular point can be examined by categorizing them with respect to eigenvalues λ. If the distinct λ_1, λ_2 are real and have the same sign, the phase trajectories lead either to the singular point ($\lambda_1, \lambda_2 < 0$) or away from it ($\lambda_1, \lambda_2 > 0$). In the former case, therefore, the singular point is a stable equilibrium, whereas in the latter case, it is unstable. Such points in the phase plane are called nodes.

When the real λ_1, λ_2 have opposite signs, the system approaches the singular point along one of the eigenvectors but moves away from it along the other. Overall, the singular point is unstable. This conclusion is valid insofar as the system may be affected by perturbations that are orthogonal to the former eigenvector, which can be expected in realistic situations. Equilibria of this type are called saddle points, in reference to how a saddle constrains the rider's motion in the forward and backward directions, but not from side to side.

The remaining realistic possibility stemming from Eq. A.5) is complex conjugate λ_1, λ_2, which imply spiral trajectories in the vicinity of the singular point. These tend to the singular point when the real part of the eigenvalues is negative, and away from it, when the real part is positive. The singular point, then, is stable or unstable, accordingly. The magnitudes of the imaginary part determine the speed of the revolutions.

Should the real part of the complex eigenvalues be equal to zero, the singularity is called a center, the trajectories around it being closed ellipses in the linear approximation. This implies neutral stability with respect to perturbations altering the magnitude of the oscillations, and a higher-order approximation may reveal the actual character of the stability. However, like the case of equal eigenvalues, to which we alluded above, this situation is not likely to arise in biological systems, except as a special case demarcating realistic regimes in the parameter space. It is worth noting however, that the classical and influential Lotka-Volterra model (see Chap. 8) is so reduced that its positive equilibrium is indeed a center point.

Whether we have a node, saddle, or spiral at the singular point is controlled by the determinant and trace of matrix A (Eq. A.3). Saddles correspond to a negative determinant, which therefore equates to instability of the singular point. When the determinant is positive, the singular point is unstable if the trace is positive and stable, if it is negative. The local system behavior depends on whether $\det A$ is above or below $(\mathrm{tr}\, A)^2/4$. Below this value (while $\det A > 0$), we have nodes, and above it, spirals.

In addition to the singular points, the limit solution may be a cycle. Existence of a stable limit cycle can be assessed by applying the Poincaré-Bendixson theorem. It states that if an unstable node or spiral-type singular point (and no other singular point) is found inside a closed boundary in the phase plane, such that all trajectories at this boundary lead inwards—which can be ascertained by examining Eq. (A.1)—then a stable limit cycle must lie between the boundary and the singular point. By applying these concepts, qualitative analysis of the system's behavior and equilibria can be rapidly carried out.

References

Guckenheimer J, Holmes PJ (1983) Nonlinear oscillations, dynamical systems and bifurcations of vector fields. Springer, Berlin

Jordan DW, Smith P (1999) Nonlinear ordinary differential equations. Oxford University Press, Oxford

Murray JD (2002) Mathematical biology I: an introduction. Springer, New York

Odell GM (1980) Qualitative theory of systems of ordinary differential equations, including phase plane analysis and the use of the Hopf bifurcation theorem. In: Segel LA (ed) Mathematical models in molecular and cellular biology. Cambridge University Press, Cambridge, pp 649–727

I. Maly, *Quantitative Elements of General Biology*,
https://doi.org/10.1007/978-3-030-79146-9

Index

A
Actin branching, 72
Action potential, 179, 180
Adaptation, 179
Adaptive selection, 18
Aedes aegypti, 14
Allolactose, 53, 54
Allopatric speciation, 173
Allopatry, 31
All-or-nothing response, 4
  canard (bifurcation), 40
  cell membrane, 38
  cooperative kinetics, 37
  cooperativity, 37
  damped oscillations, 40
  dephosphorylation, 38
  direct Hopf bifurcation, 40
  Hill coefficient, 36
  leak current, 39
  MAPK, 35, 38
  MAPKKK, 37
  membrane potential, 39
  phosphorylation, 35, 38
  plasma membrane, 35
  reversal potential, 39
  ultrasensitivity, 38
Amoeboid pseudopodia, 74
Anisotropy, 96
Annelida, 153
Apical meristem, 147, 152
Arp2/3 (protein), 62–64, 68
Array treadmilling, 62
Arthropoda, 153
Asexual and sexual reproduction, 178
Assortative mating, 28, 29
Astral microtubules, 112–117, 120–123, 125, 127–129
Astrobiology, 2
Asymmetric cell division, 99, 112, 113, 181
Asymmetric equilibrium, 102, 116–118, 120, 123, 124, 128
Asymmetry, 99–102, 114, 117, 120, 129–131, 134, 138
Autocrine EGF signaling, 87, 93
Autocrine EGFR signaling, 89
Autocrine feedback, 90
Autocrine signaling, 87, 90, 91, 94, 95
Auxin, 148, 149, 152
Auxin-controlled growth, 150
Axial symmetry, 115, 128
Axisymmetric perturbations, 128
Axons, 38

B
Bacteria, 7
Bayesian inference, 180
Beam equation, 103, 105
Bending elasticity, 101, 113, 114
Bifurcation, 12, 40, 41, 50, 51, 68, 70, 73, 91, 92
Bilateria, 153
Biochemical oscillations, 42
Biogeography, 3
Biological control mechanisms, 177
Biomass, 3, 5, 10, 164
Biomass growth, dynamic systems
  bacteria, 7
  biomass steady state, 10
  chaos, 10–14

I. Maly, *Quantitative Elements of General Biology*,
https://doi.org/10.1007/978-3-030-79146-9

Biomass growth, dynamic systems (*cont.*)
competition (ecological), 8
continuous-time univariate models, 9
cycle, 12–14
equilibrium, 9
generations, 8
growth rate function, 9
homogeneous models, 8
hysteresis effects, 10
immigration, 10
measurement-related noise, 14
oscillations, 10–14
period-doubling, 13
perturbations, 7, 9, 11
population, 7, 8
reproduction function, 12
second-order rate, 8
stable equilibrium, 9
stochastic process, 14
third-order rate constant, 8
unstable equilibrium, 9, 11
Biomass kinetics, 177, 179
Biomechanics, 181
Blastomeres, 113
Blood, 45–47
Blowfly, 14
Boltzmann factor, 70
Brain stem, 43
Brain stem control, breathing, 42, 44
Branching-capping operator, 68, 74
Breathing control model, 43
Breeding season, 10
Brownian ratchet model, 62, 70
Buckling, 105, 107, 109
Buckling force, 113, 116–118, 123
Budworm outbreak model, 10
Burger's equation, 156

C
Caenorhabditis (roundworm), 112–114, 128–130
Calcium system model, 50
Canard (bifurcation), 40
Capacitance density, 38, 43
Capping protein, 63, 64, 68
Cartesian coordinate system, 114
Catabolite activator protein (CAP), 53
Caudal bud, 154
Cell biology, 1, 3
Cell boundary, 99–103, 106–107, 109, 112–114, 116, 118, 125, 129–131, 135
Cell boundary deformation, 62
Cell cortex, 62, 70, 72
Cell cycle, 181
Cell division, 5, 112–129
asymmetric, 99, 112, 113
equal, 114
oriented, 130
symmetric, 112, 113
unequal, 114
Cell growth, 181
Cell interaction, 129–139
Cell lines, 112
Cell locomotion, 62
Cell mechanics, 177, 181
Cell membrane, 38, 73, 89, 90
Cell morphogenesis, 63
Cell morphologies, 74
Cell motility, 61
Cell polarity, 129
autocrine EGF signaling, 87
autocrine signaling, 90, 95
bifurcation, 91
cell polarization, 87
cell signaling, 95
dephosphorylation, MAP, 90
diffusion constant, 93
diffusion-like mechanisms, 87
EGF, 87
EGFR, 88, 89
ERK-SOS negative feedback, 88
extracellular diffusion constant, 94
Grb2, 87, 88, 95
intracellular signaling, 92
ligand, 90
MAPK, 94
morphogenetic gradients, 96
motility, 87
negative feedback, 94
partial differential equations, 88
plasma membrane, 92
positive feedback, 91, 94
Ras, 95
self-assembly, 87
signal transduction, 90, 94
SOS (protein), 87, 88, 95
supramolecular complexes, 87
tissue anisotropy, 96
Cell polarization, 87
Cell signaling molecules, 181
Cellular organization, 95
Central nervous system, 57
Centrosome, 75, 76, 86, 111
and cell center, 107, 108
cell-cell interface, 130
directions of buckling, 109
displacement, 105

displacing force, 109
equilibrium centrosome position, 111
equilibrium distance, 102, 108
interphase, 115
lymphocyte, 133, 138
mechanical model, 137
microtubules, 100
oscillations, 132, 136
reorientation, 132
spindle-pole, 113
synapse, 136
T cell centrosome, 131
trajectories, 134, 137
Centrosome migration, 137
Centrosome positioning, 100–102, 130
Centrosome-free cell fragment, 75, 79
Chaos, 4, 10–14
Chemoreceptors, 45
Chordata, 153
CO_2 chemoreceptors, 47
Coevolution, 5
Coevolutionary avalanches, 166
Colorado potato beetle, 14
Community matrix, 165
Comparative biology, 1
Competition (ecological), 161–168, 170
Competitive niche, 163
Conductivity, 39
Continuous-time biomass growth, 178
Cooperative enzymes, 36
Cooperative kinetics, 37
Cooperativity, 37, 53
Cortex
actin branching, 72
actin filament dynamics, 63
actin network, 74
bifurcation, 68
Brownian ratchet model, 62, 70
cell locomotion, 62
cortical actin cytoskeleton, 62
cortical actin network, 62
cortical actin self-organization, 68
cytoskeleton dynamics, 61
dendritic nucleation, 64
eigenfunction, 68, 74
Fisherian fitness, 66
lamellipodium, 64, 67, 70
molecular quasi-species equation, 74
Monte Carlo method, 73
motile cells, 70
motile metazoan cells, 72
mutation, 67, 68
natural selection, 61
nucleation, 64
organelles, 61
plasma membrane, 63
polymerization, 64
quasi-species, 74
reproduction, 65
retrograde flow, 62
self-assembly, 63
spatial-kinetic theory, 70
supramolecular structure, 61
tissue morphogenesis, 71
Cortical actin cytoskeleton, 62
Cortical actin filament, 65
Cortical actin network, 62
Cortical actin self-organization, 68
Cortical actomyosin, 100
Coupling strengths, 167
Cyclic adenosine monophosphate (cAMP), 52
Cyclic dynamics, 179
Cyclicity, 180
Cyclostomata, 42
Cytoplasm, 76, 78, 180
Cytoskeletal fibers, 180
Cytoskeleton, 4, 99–114, 116, 118, 123, 128–131, 135, 138, 181
acentrosomal radial array, 75
centrosome, 75, 86
centrosome-free cell fragment, 79
cytoplasm, 76, 78
cytoplasm self-centering process, 76
dynein, 84
dynein molecular motors, 76
Eukaryota, 74
melanophore, 80
microtubule and nucleator distributions, 82
microtubule array self-organization, 85
microtubule nucleators, 81
microtubule polymerization, 77
microtubule treadmilling, 81
microtubules, 77
molecular motors, 75
nucleation, 75
organelles, 75
partial differential equations, 85
radial microtubule array, 75
radial organization, 81
treadmilling microtubules, 76, 79
treadmilling polymerization, 75
tubulin polymerization, 77
Cytosolic calcium concentration, 51

D

Damped oscillations, 40, 50
Darwinian dynamics, 63

Dendritic nucleation, 62–64
Dephosphorylation, 37, 38, 41, 88, 90
Depolymerization, 62, 77, 85
Deuterostomia, 153
Diaphragm, 45, 47
Diffusion, 87, 88, 90, 92
Diffusion coefficient, 149, 171
Diffusion constant, 93
Direct Hopf bifurcation, 40
Dissipative oscillations, 135, 139
Douady-Couder style models, 151
Drosophila, 112
Dynamical system, 2–5, 177, 178, 180–183
Dynein, 75, 81, 84, 100, 130, 131, 135

E
Earth, 180
Ecdysozoa, 153
Echinodermata, 153
Ecological relationship, 166, 170
EGF autocrine network, 88
EGF receptor (EGFR), 55, 88–90
Eigenfunction, 68, 74
Eigenvalue, 67, 68, 165
Eigenvectors, 165
Embryo segmentation, 182
Embryology, 1, 153
Endocytosis, 88
Energy-dissipating biochemistry, 135
Energy-dissipating process, 106
Energy minimum, 101, 102, 109
Epidermal growth factor (EGF), 55, 56, 87, 88
Epidermis, 147, 148
Epithelioid cells, 100, 101
Epithelium, 130
ERK (kinase), 87–89, 91, 94
Erk2, 35
ERK-dependent SOS phosphorylation, 94, 95
Escherichia coli, 52
Eukaryota, 74
Eukaryotic cell signaling, 55
Euler problem, 104
Euler-Lagrange solution, 104
Evolutionarily stable strategies, 168
Extracellular diffusion constant, 94
Extraterrestrial life, 2

F
Feedback, 68, 88, 90, 91, 94, 95
Fibroblast growth factor (FGF), 146, 154
Fisher (Ronald), 66
Fisherian fitness, 66
Force-balance models, 113
Friction, 111

G
Galactosidase, 53
Gametes, 22
Gas exchange, 45
Gene expression, 180
Gene interaction network, 153, 157
General biology
  comparative biology, 1
  dynamical systems, 2, 3
  ecology, 1
  evolution, 1
  extraterrestrial life, 2
  generalizable biology, 2
  goal, 1
  life, 2
  molecular biology, 1
  physiology, 1
  quantitative approach, 1
  self-organization science, 1
  system-level properties, 1
Generalizable biology, 2
Genetic cluster, 17
Genetic models, 166
Genetic oscillations, 154
Genetics, 1
Genetic-spatial evolutionary mechanisms, 179
Genome space, 5, 27, 166, 173
Genomics, 1
Genotype clusters, 166
Genotypes, 18–20
Genotype space, 161, 166–169
Geographic isolation, 173
Geographical (planetographical) distributions, 170
Geographical (planetographical) separation, 173
Germ line, 112, 130
Glucose, 52, 53
Glycolysis, 42
Grb2, 56, 87–89, 94, 95
Growth factor receptors, 42
Growth tensor, 150
Gymnocorymbus ternetzi, 70
Gyrations, 132

H
Haploid genome, 18
HeLa cultured cells, 113
Hemoglobin, 36

Heritable cell structure, 180
Hermaphrodites, 8
Heterozygosity, 26
Heterozygotes, 21, 25
Higgs-Derrida model, 168, 173
Hill coefficient, 36, 37
Hill equation, 36, 37
History-dependence, 102
Hodgkin-Huxley equation, 43
Hodgkin-Huxley model, 38, 40, 43, 45
Homeostasis, 35, 179, 180
Homozygosity, 28
Homozygotes, 25
Hookean elasticity, 118
Horizontal gene transfer, 24
Hybrids, 25
Hydrolysis, 54
Hypercapnia, 47
Hysteresis, 52, 55, 57, 180

I
Immune cells, 100
Immune system, 100
Immunological synapse, 130–133, 138
Inbreeding coefficient, 28
Insecta, 153
Inspired volume, 45
Instability, 101–103, 117, 120, 123, 125–128, 135, 138, 139
Interaction matrix, 163
Interphase, 5, 112, 114, 115, 123, 128, 129
Inter-species dynamics, 14
Intracellular kinase, 179
Intracellular signaling, 90, 92
Inverted Hopf bifurcation, 40, 41
Irreversibility, 52, 99–102, 104–107, 109, 112, 131, 138

J
Jacobian matrix, 171

L
lac operon, 52–55
Lactose, 52–55
Lactose metabolism, 52
Lactose permease, 52
Lamellipodium, 64, 67–71
Leaf primordia, 143
Leak current, 39, 43
Leaves, 143, 144, 147
Leech (*Helobdella*), 112
Leptinotarsa decemlineata, 14
Ligand, 87, 90
Linear approximation, 186
Living matter, 4, 7, 24
Locomotion, 64, 87
Lophotrochozoa, 153
Lotka (Alfred), 5
Lotka's kinetic model, 164
Lotka-Volterra approach, 172
Lotka-Volterra equations, 164
Lotka-Volterra model, 165, 187
Lotka-Volterra predator-prey system, 165
Lucilia cuprina, 14
Lung, 45
Lymphocytes, 100, 101, 107, 132, 133, 137, 138

M
Machine learning, 1
MAPK kinase cascade, 55
Master-sequence model, 19
Mathematical methods, 185
Mechanics
  bundling, 131
  centrosome, 100, 137
  cytoskeleton, 100, 102, 138
  hysteretic loop, 135
  instability, 138
  manifestations of life, 111
  separate poles, 123
  systems biomechanics, 99
Mechanochemical formalism, 182
Mek1, 35
Melanophore, 80, 81
Melanosome, 81
Membrane potential, 38–40, 43
Mendel (Gregor), 1
Meristem, 147–149
Mesoderm, 154, 156
Metabolism, 1, 181
Metapopulation, 173
Metastable equilibrium, 106
Michaelis-Menten constant, 41
Michaelis-Menten equation, 36
Michaelis-Menten kinetics, 41
Microphotographs, 115
Microtubule cytoskeleton, 99–102, 104, 106–114, 118, 128–131, 181
Microtubule nucleation, 80–81
Microtubule polymerization, 100
Microtubule treadmilling, 76, 81, 86

Microtubules
aster, 100, 111, 131, 134
astral, 113–117, 120–123, 127–129
bending elasticity, 101
bending rigidity, 117
buckling forces, 117
buckling forms, 105
bundling, 131
cell boundary, 113
centrosome, 100
convergence, 130
cytoskeleton, 99–102, 104, 106, 108–114, 118, 128–131
deformation, 113, 114, 135
elasticity, 135
equilibrium forms, 106
flat spatial domain, 109
length, 131
length and cell radius, 102, 107
length distributions, 125
lift-off, 137
long and short, 125
metastable, 105
nonlinearity, 109
parameters, 114
polymerization dynamics, 100
single, 102
spindle-pole centrosomes, 113
trailing, 131
types, 104
unstrained directions, 109
Migration, 9, 173
Misoriented cell-cell conjugate, 131
Mitogen activated protein kinase (MAPK), 35, 87, 88, 90, 94
Hill equation, 36
MAPKK, 38
negative feedback, 41
oscillations, 41, 42
phosphorylation, 35, 36, 38, 41
SOS, 42
Mitosis, 35, 115, 126
Mitotic spindle, 112, 115, 127, 130
Molecular biology, 1
Molecular motors, 75, 76, 79, 84, 100, 101, 113, 130, 131
Molecular paleontology data, 30
Molecular quasi-species equation, 74
Molecular self-organization, 4
Mollusca, 153
Monomorphic equilibrium, 25
Monte Carlo method, 73
Morphogenesis, 5, 62, 71, 99, 112, 113, 123, 125, 139, 181
Morphogenetic gradients, 96
Mosquito, 14
Motility, 61, 83, 87, 99, 181
Multicellular domain, 182
Multicellular morphogenesis, 181
phyllotaxis (*see* Phyllotaxis)
segmentation (*see* Segmentation, multicellular morphogenesis)
Turing mechanisms (*see* Turing mechanisms)
Multiperiodic motion, 129–139
Multiperiodicity, 4
Multistability, 100, 129
Muscles, 112, 128, 177
Mutation, 63, 64, 67, 68
Mutation-selection balance, 4
Mutation-selection equilibria, 177, 179
Mutualism, 5, 161, 182

N
Negative feedback, 88
Neuroanatomy, 42
Neuroblasts, 112
Neuronal physiology, 40
Neurons, 40, 43, 47, 57
Neurophysiology, 42
Neutral equilibrium, 117
Neutral evolution, 177
Neutral speciation, 173
Neutral stability, 165
Neutral sympatric speciation, 167
alleles, 26
asexual case, 24, 27
asexual populations, 27
assortative mating, 28, 29
clonal mechanism, 24
cycles, 25
ecological and geographic niches, 24
evolution, 26
extinction, 28
genealogical branch, 31
generations, 28
genome, 26
genomic space, 27
Higgs-Derrida model, 31
homozygosity, 28
inbreeding coefficient, 28
individual-based stochastic simulations, 25
local stability, 25
monomorphic population, 25
multi-locus systems and reproductive isolation, 26
non-overlapping generations, 24
population bottleneck, 25, 28, 30, 32

punctuated equilibria, 31
radiation (evolutionary), 30
randomly mating population, 27
random-mating model, 27
reproductive isolation, 24, 32
self-organized species, 28
sexual model, 27
single-locus models, 24, 32
steady-state population, 28
steady-state solutions, 25
Nonaxisymmetric equilibrium, 128
Nonaxisymmetric perturbations, 128
Non-competitive inhibition, 41
Nondimensionalization, 162
Non-equilibrium polymerization, 62
Nonlinear dynamics, 4
Non-overlapping generations, 11, 14, 178
Notch (protein), 154
Nucleation, 62, 64, 75–77, 81, 83
Nucleators, 86
Nucleus, 100, 131, 132, 138

O
Oocytes, 112, 128
Operator (sequence), 53, 54
Operon, 53, 54
Organelles, 61, 75, 79, 100
Orixate phyllotaxis, 147, 152
Oscillation cycle, 134
Oscillations, 4, 10–14, 154, 156, 158
Oscillatory behavior, 177
Oscillatory control processes
bifurcation, 50
biochemical oscillations, 42
brain stem control, breathing, 42
calcium system model, 50
cycle, neural activity, 43
cytosol, 49, 50
gas exchange, 45
Hodgkin-Huxley equation, 43
Hodgkin-Huxley model, 43, 45
inter-organellar calcium dynamics, 52
intracellular calcium system, 48
kinetics, 50
MAPK, 41
mitochondrial calcium concentration, 49
negative feedback, 41
neuron populations, 46
organelles, 49
plasma membrane, 47
positive feedback, 42
receptor channel, 50
steady-state oscillations, 41

P
Parapatric speciation, 173, 179
Partial differential equations, 76, 85, 88, 90
Perturbations shift, 165
Petals, 147
Phase plane analysis, 185–187
Phase trajectories, 164
Phosphatases, 36, 38
Phosphorylation, 35–38, 41, 55, 56
Phyllotaxis, 5
auxin, 148, 149
dicussate, 147
diffusion coefficient, 149
Douady-Couder style models, 151
epidermis, 147
generative model, 152
growth tensor, 150
inhibitory-field model parameters, 152
instability, 149
mechanical and biochemical processes, 148
mechanochemical model, 150
meristem, 147, 148
orixate, 147, 152
petals, 147
phyllotactic patterns, 151
Poisson ratio, 149, 150
sepals, 147
stress tensor, 150
tunica, 148, 150
Turing theory, 148
vascular plants, 147
Young's modulus, 149
Physiological adaptation, 4
Physiology, 1, 3, 42
Plasma membrane, 35, 38, 41, 43, 47, 48, 50, 55, 62–64, 71, 89–92, 95
Poincaré-Bendixson theorem, 187
Poisson ratio, 149, 150
Pollen tubes, 72, 74
Polymerization, 61–64, 70, 75–77
Population bottleneck, 25, 28, 30, 32
Positive feedback, 53, 56, 88, 93
Post-mating isolation, 26
Potassium channels, 38
Potassium rectifier current, 43
Predation, 182
Predator-prey model, 172
Predator-prey relationship, 161, 173
Predator-prey system, 172
Presomitic mesoderm, 154, 156
Priapulida, 153
Pronucleus, 134
Punctuated equilibria, 31
Punctuated equilibrium, 5

Q
Quasi-species, 19, 74
Quasi-species theory, 4, 178, 181
Quasi-stable states, 168

R
Radiation (evolutionary), 28
Raf, 55, 56, 88
Random-mating model, 27
Ras, 41, 95
Reaction-diffusion mechanism, 87
Receptor channel, 39, 50
Reflection symmetry, 109, 123
Reproduction, 61, 63–65, 74
Reproductive isolation, 17, 24, 26, 28, 32
Retinoic acid, 154
Retrograde flow, 62, 70
Retrotrapezoid nucleus (RTN), 43
Reversal potential, 39, 46
Reynolds number conditions, 134
RNA polymerase, 53
Root hairs, 72, 74
Rotational symmetry, 109

S
Saddle points, 186
Scalar biomass models, 178
Scar (protein), 62
Segmentation, 5
Segmentation, multicellular morphogenesis
  Arthropoda, 153
  Bilateria, 153
  Burger's equation, 156
  Chordata, 153
  Deuterostomia, 153
  developmental organization, 153
  Ecdysozoa, 153
  embryology, 153
  evolutionary simulations, 153
  FGF, 154
  gene interaction network, 153, 157
  genetic oscillations, 154
  Lophotrochozoa, 153
  Mollusca, 153
  morphogenetic molecule, 153
  Notch (protein), 154
  oscillations, 154, 156
  presomitic mesoderm, 154
  Priapulida, 153
  rostral somites, 154
  simultaneous segmentation, 153, 154, 158
  somites, 154
  somitogenesis, 157
  Wnt-Notch-FGF system, 155
Selection-mutation equilibria, 178
Selection-mutation equilibrium, species
  adaptive selection, 18
  alleles, 19, 23
  allele-type frequencies, 22
  Darwin (Charles), 17
  diploid frequencies, 22
  diploid genetic distributions, 20, 21
  diploid genotypes, 23
  diploid populations, 20
  diploids, 21
  diploid-stage genotype
    frequencies, 22
  direct replication, 21
  error threshold, 18–21, 23
  fitness alleles, 18, 20, 22
  fitness landscape, 19, 21
  gametes, 22
  genetic cluster, 17
  genetic distributions, 23
  genome types, 21
  genome-wide mutation rate, 20
  genotype, 18–20
  genotype frequencies, 20
  heterozygotes, 21
  heterozygous locus, 20
  locus, 18, 20
  master-sequence model, 19
  multi-locus haploid genome, 18
  multiplicative landscape, 21
  mutant locus, 20
  population, 24
  reproductive isolation, 17
  sexual model, 22
  steady-state frequency, 24
  Wallace (Alfred), 17
Selection-mutation models, 178
Self-assembly, 63, 71, 87
Self-fertilizing hermaphrodites, 8
Self-organization, 5, 99, 129, 139
Self-organization science, 1
Self-organization, cell
  cell polarity (*see* Cell polarity)
  cortex (*see* Cortex)
  cytoskeleton (*see* Cytoskeleton)
Self-organized criticality, 166, 182
Sensory physiology, 57
Sepals, 147
Shc, 56
Shc-Grb2-SOS protein complex, 55
Signal processing, 4
Signal transduction, 42, 87, 88, 90, 91, 94, 95

Signaling network, 90, 91
Simultaneous segmentation, 153, 154, 158
Single-species biomass kinetics, 3
Singular point, 164, 165, 186, 187
Soil, 170
Somites, 154
Somitogenesis, 157
SOS (protein), 41, 42, 56, 87–89, 94, 95
Spatial self-organization, 181
Spatially distributed mutualistic system, 172
Speciation, 3–5, 177, 179
Species, 3–5
Species and speciation
- neutral sympatric speciation (*see* Neutral sympatric speciation)
- selection-mutation equilibria (*see* Selection-mutation equilibrium, species)

Species density patterns, 172
Species interaction, 177, 182
Species interaction dynamics
- biomass, 161
- coexistence and mutual exclusion, 161
- competitive niche, 163
- interaction matrix, 163
- mutualism, 161, 162
- phase plane analysis, 162
- spatial effects, 170–174
- system's phase space, 164
- temporal patterns, 164–170

Spindle poles, 112–115, 118, 128
Spindle positioning, 113, 114, 116, 120, 129, 130
Spiral-type singular point, 187
Stability, 4, 35, 50
Stability analysis, 102, 114
Static equilibrium, 109, 138
Steady-state density functions, 68
Stochastic process, 14
Stochasticity, 102, 134
Stress tensor, 150
Supramolecular complexes, 70, 87
Switches, 52–57
Symmetric and asymmetric division, 177
Symmetric cell division, 112, 113
Symmetric equilibrium, 101, 116–118, 120, 124, 127, 129
Symmetry-breaking, 99–107, 109, 111, 112, 129
Sympatric speciation, 4
Sympatry, 24
Systems biomechanics, 99, 103, 120, 139

T

Tangled-nature model, 166, 168, 169
Taxonomy, 24, 27
Temporal oscillations, 157
Temporal patterns, 164–170
Tentacles, 143
Thermodynamics, 111, 135
Tissue anisotropy, 96
T-killer lymphocyte, 129
Topopatric mechanisms, 174
Traveling waves, 57
Treadmilling (kinetics), 75, 76, 79
Treadmilling microtubules, 76, 80, 82
Tubulin polymerization, 77, 79, 101
Tunica, 148–150
Turing (Alan), 5
Turing diffusional instability, 171
Turing mechanisms
- biology, 143
- *Bmp* genes, 146
- diffusion, 145
- diffusion-driven instability, 144, 147
- *FGF*, 146
- homogeneous molecular system, 143
- homogeneous steady state, 144
- *Hox* gene, 146
- hydrozoan tentacles, 143
- instability, 143, 146
- leaf primordia, 143
- leaves, 143
- morphological elements, 143
- multicellular morphogenesis, 143
- organ-level morphogenesis, 146
- perturbations, 144, 145
- phyllotaxis, 143
- rugae formation, 146
- segmentation, 143
- Shh pattern, 146
- Wnt (protein), 146

U

Ultrasensitivity, 38
Unequal cell division, 112, 114

V

Vascular plants, 5, 147
Vertebrata, 153
Viscosity, 111, 134, 135, 137
Viscous drag, 135, 137
Volterra (Vito), 5
Volterra model, 164
Volterra's ecological investigation, 164

W

Wallace effect, 24
WASP (protein), 62
Whole-cell movement, 131
Whorled phyllotaxis, 143
Wiskott-Aldrich syndrome protein
(WASP), 62
Wnt-Notch-FGF
system, 155
Wnt (protein), 146, 154
Wound closure, 100
Wound healing, 57

X

Xenopus, 38, 70

Y

Young's modulus, 149